D1084188

Texts and Monographs in Computer Science

Texts and Monographs in Computer Science

Edsger W. Dijkstra Carel S. Scholten

Predicate Calculus and Program Semantics

Springer-Verlag New York Berlin Heidelberg
London Paris Tokyo Hong Kong

Edsger W. Dijkstra
University of Texas at Austin
Austin, TX 78712-1111
USA

Carel S. Scholten
Klein Paradys 4
7361 TD Beekbergen
The Netherlands

Series Editor

David Gries
Department of Computer Science
Cornell University
Ithaca, NY 14853
USA

Library of Congress Cataloging-in-Publication Data
Dijkstra, Edsger Wybe.
  Predicate calculus and program semantics / Edsger W. Dijkstra,
Carel S. Scholten.
    p.   cm. — (Texts and monographs in computer science)
  ISBN 0-387-96957-8 (alk. paper)
  1. Predicate calculus.   2. Programming languages (Electronic
computers)—Semantics.   I. Scholten, Carel S.   II. Title.
III. Series.
QA9.35.D55 1989
511.3—dc20   89-11540

Index prepared by Jim Farned of The Information Bank, Summerland, California.

Typeset by Associated Publishing Services, Ltd., United Kingdom.
Printed and bound by R.R. Donnelley & Sons, Harrisonburg, Virginia.
Printed in the United States of America.

9 8 7 6 5 4 3 2 1

ISBN 0-387-96957-8 Springer-Verlag New York Berlin Heidelberg
ISBN 3-540-96957-8 Springer-Verlag Berlin Heidelberg New York

CHAPTER 0
Preface

This booklet presents a reasonably self-contained theory of predicate transformer semantics. Predicate transformers were introduced by one of us (EWD) as a means for defining programming language semantics in a way that would directly support the systematic development of programs from their formal specifications.

They met their original goal, but as time went on and program derivation became a more and more formal activity, their informal introduction and the fact that many of their properties had never been proved became more and more unsatisfactory. And so did the original exclusion of unbounded nondeterminacy. In 1982 we started to remedy these shortcomings. This little monograph is a result of that work.

A possible —and even likely— criticism is that anyone sufficiently versed in lattice theory can easily derive all of our results himself. That criticism would be correct but somewhat beside the point. The first remark is that the average book on lattice theory is several times fatter (and probably less self-contained) than this booklet. The second remark is that the predicate transformer semantics provided only one of the reasons for going through the pains of publication.

Probably conditioned by years of formal program derivation, we approached the task of designing the theory we needed as an exercise in formal mathematics, little suspecting that we were heading for a few of the most pleasant surprises in our professional lives. After a few notational adaptations of the predicate calculus —so as to make it more geared to our manipulative needs— and the adoption of a carefully designed, strict format for our proofs, we found ourselves in possession of a tool that surpassed our wildest expectations. As we got used to it, it became an absolute delight to work with.

The first pleasant —and very encouraging!— experience was the killing of the myth that formal proofs are of necessity long, tedious, laborious, error-prone, and what-have-you. On the contrary, our proofs turned out to be short and simple to check, carried out —as they are— in straightforward manipulations from a modest repertoire.

For quite a while, each new and surprisingly effective proof was a source of delight and excitement, although it was intellectually not fully satisfactory that each of them was a well-crafted, but isolated, piece of ingenuity. We had our second very pleasant surprise with the development of heuristics from which most of the arguments emerged most smoothly (almost according to the principle "There is really only one thing one can do."). The heuristics turned the design of beautiful, formal proofs into an eminently teachable subject.

This experience opened our eyes for further virtues of presenting formal proofs in a strict format. Proofs thus become formal texts meeting precise consistency criteria. The advantage of this transition is traditionally viewed as a gain in trustworthiness: given the formal text, there need not be an argument whether the proposed proof is a proof or not since it can be checked objectively whether the text meets the criteria. (This is the aspect stressed in mechanical proof verification.) A further advantage of the transition is that it enables a meaningful comparison between alternative proofs of the same theorem, viz., by comparing the texts. The final advantage of the transition is the much greater homogeneity it introduces into the task of proof design, a task that becomes a challenge in text manipulation that is more or less independent of what the theorem was about. It is this greater homogeneity that opens the way for heuristics that are more generally applicable.

In the course of the process we profoundly changed our ways of doing mathematics, of teaching it, and of teaching how to do it. Consequently, this booklet is probably as much about our new appreciation of the mathematical activity as it is about programming language semantics. It is certainly this second aspect of the booklet that has induced us to go through the aforementioned pains of publication. It has taken us some time to muster the courage to admit this, for a revision of mathematical methodology seemed at first a rather presumptuous undertaking. As time went on, however, we were forced to conclude that the formal techniques we were trying out had never been given a fair chance, the evidence being the repeated observation that most mathematicians lack the tools needed for the skilful manipulation of logical formulae. We gave them a fair chance; the reader is invited to share our delight.

*   *   *

Before the reader embarks on the study of the material proper of this booklet, we would like to give him some advice on reading it.

The most important recommendation is to remember all the time that this is *not* a treatise on logic, or on the foundations of mathematics. This warning is all the more necessary since superficially it might look like one: it has logical symbols all over the place and starts with a long introduction about notation. The frequent occurrence of logical symbols has a simple, pragmatic explanation: we use them so extensively because they are so well-suited for our job. The long introduction on notation has an equally simple, pragmatic explanation: some new notation had to be introduced and, more importantly, existing notations had to be adapted to our manipulative needs. Our free use of the logical connectives before their formal introduction stresses once more that this is not a book on logic. In short, the logician is perfectly free to be taken aback by our naive refusal to make some of his cherished distinctions.

Our second recommendation to the reader is to approach this little monograph with an open mind and not to get alarmed whenever it deviates from the traditions in which he happens to have been educated. In particular, he should not equate convenience with convention. Doing arithmetic in Arabic numerals is objectively simpler —i.e., more convenient— than doing it in Roman numerals. The transition from verbal arguments appealing to "intuition" or "common sense" to calculational reasoning admits also in that area an equally objective notion of simplicity —i.e., of convenience— . We know from experience that for readers from some cultures it will be hard to accept that we leave all sorts of (philosophical or psychological) questions unanswered; the only answer we can offer is that we are from a pragmatic culture that deals with such questions by not raising them.

Our third recommendation to the reader is really a request, viz., to honour this booklet's brevity by reading it slowly. Our texts have a tendency of being misleadingly smooth and unusually compact. When, at the end, you wonder "Was this all?", we shall answer "Yes, this was all. And we hope you travelled long and far.".

*   *   *

The above was written a year ago, before we had started on our manuscript. It reflects our expectations. Now, 12 months and 420 handwritten pages later, we can look back on what we have actually done besides breaking in a new fountain pen. Needless to say, the major discrepancy between dream and reality has been the size: had we foreseen a 420-page manuscript, we would not have referred to a "booklet". Our only excuse is that, at the time, we had not firmly decided yet to include the material now covered in the last three chapters.

The trouble with writing a book is that it takes time and that, consequently, at the time of completion, authors are older —and perhaps wiser— than when they started. Our fate has been no exception: there are several things we would have done differently had we written them a year later. (For instance, concerns about punctuality would have been more concentrated and our treatment of the implication in Chapter 5 might have relied less heavily on the "Little Theory".) With the exception of a complete rewriting of Chapter 1, which at the end was no longer a proper introduction to the rest of the book, we have, however, abstained from any major overhauls of the manuscript. They would not have solved the problem that authors may continue to grow and that, consequently, texts have an unavoidable tendency to be more or less dated. It is in this connection that we would like to include a general disclaimer: though our enthusiasm might sometimes suggest differently, we nowhere pretend that our work leaves no room for improvement, simplification, or meaningful generalization. (We did, for instance, not explore what meaningful partial orders on programs could be introduced.)

Apart from exceeding the originally envisaged size, we have remained faithful to our original intentions, in particular to our intention of writing a monograph reflecting the current state of our art. Though we have successfully covered most of its material in graduate courses at both sides of the Atlantic Ocean, this book should not be regarded (or judged) as a textbook, because it was not intended that way. (Hence, for instance, the total absence of exercises.) There were several reasons for not wishing to write a textbook. A practical reason was that different educational systems promote totally different perceptions of student needs to be catered for in an "acceptable" textbook, and that we did not want to waste our time trying to meet all sorts of conflicting requirements. A more fundamental reason was that we think the whole notion of "a textbook tailored to student needs" certainly at graduate level much too condescending. At that stage there is nothing wrong with confronting students with material for which they have not been adequately prepared, for how else are they going to discover and to learn how to cope with the unavoidable gaps in their education? So, though we know that this book can provide the underlying material for a fascinating and highly instructive course, it is *not* a textbook and its "target audience" is just "To whom it may concern".

We have, of course, our idea about whom it concerns: the mathematically inclined computing scientists and the mathematicians with methodological and formal interests. We most sincerely hope to reach them, to thrill them, and to inspire them to improve their own work or ours. Honesty compels us to add to this wish that there is one possible —and, alas, likely— "improvement" we are not waiting for, viz., the translation of our theory into set-theoretical terminology —by interpreting predicates as characteristic functions of subsets of states— so as to make it all more familiar. Little is so regrettable as to see one's work "improved upon" by the introduction of traditional complications one has been very careful to avoid. Hybrid arguments, partly conducted in terms of a formal system and partly conducted in terms of a specific model for that formal system, present a typical example of such confusing complications. In this connection we would like to stress that the existence of individual machine states enters the picture only when our theory is applied to program semantics, i.e., to an environment in which the individual machine state is a meaniyful concept; the theory itself does not need a postulate of the existence of individual states and, therefore, should not be cluttered by their introduction.

*   *   *

It is a pleasure to mention here our great appreciation for our former employers, Burroughs Corporation and N.V. Philips, respectively, which loyally supported us when we embarked in 1982 on the investigations reported in this volume.

More profound gratitude than to anyone else is due to W. H. J. Feijen and A. J. M. van Gasteren who were at that time close collaborators of one of us (EWD). They did not only witness from close quarters our first formal efforts at establishing a theory of predicate transformer semantics, they can trace their contributions and their influence all through this book. Feijen is essentially the inventor of our proof format; he took the decision to write the hint between the formulae connected by it and he was the one who insisted on the discipline of allotting at least a full line to each hint. (He probably realized that this insistence was necessary for establishing a tradition in which the hints given would be as explicit as he wanted them to be.) Van Gasteren provided the rationale for this invention (and also for the invention of the square brackets to denote the "everywhere" operator): her earlier explorations had convinced us that the type of brevity thus obtained is indispensable. Later she insisted on exorcizing mathematical rabbits —pulled out of a hat— and provided the "follows from" operator as one of the means to that end.

Furthermore, we thank all colleagues, students, and members of the Tuesday Afternoon Clubs in Eindhoven and Austin, whose reactions to the material shown have helped in shaping this book. We are particularly grateful for the more detailed comments that Jayadev Misra and Lincoln A. Wallen gave on the almost final version of the manuscript; the decision to rewrite Chapter 1 has been triggered by their comments.

Finally, we express our gratitude to W. H. J. Feijen, David Gries, and Gerhard Rossbach of Springer-Verlag, New York. In their offers of assistance

in the final stages of book production, each of them has gone beyond the call
of duty.

July 1988
Austin, TX, USA   Edsger W. Dijkstra
Beekbergen, The Netherlands   Carel S. Scholten

Contents

CHAPTER 1

On structures

The proofs in this book are much more calculational than we were used to only a few years ago. As we shall explain later, the theorems are (or could be) formulated as boolean expressions, for which, in principle, *true* and *false* are the possible values; the proofs consist in calculations evaluating these boolean expressions to *true* . We shall return to this later, focussing, for the time being, our attention on some of the notational consequences of this approach.

The advantages of such a calculational style are a fairly homogeneous proof format and the possibility of obtaining brevity without committing the "sin of omission", i.e., making such big leaps in the argument that the reader is left wondering how to justify them. In fact, all our steps are simple and they are taken from a repertoire so small that the reader can familiarize himself with it as we go along. We could, however, harvest these advantages of calculation only by adoption of carefully chosen notational conventions that tailored our formulae to our manipulative needs. (Among those needs we mention nonambiguity, brevity, and not being forced to make needless distinctions.)

One of our notational conventions may strike the reader as a gross overloading of all sorts of familiar operators: for instance, we apply the operators from familiar two-valued logic to operands that in some admissible models may take on uncountably many distinct values. The justification for such a notational convention is a major purpose of this introductory chapter; we feel that we owe the reader such a justification, all the more so because in the world of programming the dangers of overloading are well known.

We can get some inspiration —and, if we feel the need for it, even some reassurance— from the field of physics. Every classical physicist, for instance, is thoroughly familiar, be it in his own way, with the notion of a vector in three-dimensional Euclidean space, independently of the question of whether the vector is a displacement, a velocity, a force, an acceleration or a component of the electromagnetic field. Also, he is equally familiar with the sum $v + w$ of two vectors v and w . But that sum raises a question, in particular if one adopts the view —as some physicists do— that the variables stand for the physical quantities themselves and not for their measure in some units. The question is, how many different vector additions are used by the physicist: is the sum of two velocities the same sort of sum as the sum of two forces? Well, the answer seems negative in the sense that no physicist that is well in his mind will ever add a velocity to a force.

Given the fact that in some way we can distinguish those different sorts of additions, we could feel tempted or intellectually obliged to introduce as many different addition symbols as we can distinguish additions, say $+_v$ for the addition of velocities, $+_a$ for the addition of accelerations, etc. In a purist way, this would be very correct, but we all know that the physical community has decided against it: it has decided that a single symbol for addition will do.

When challenged to defend that decision, the physicist will give the following reasons. Firstly, the purist convention would complicate manipulation: the single rule that differentiation distributes over addition, i.e.,

$$\frac{d}{dt}(v + w) = \frac{dv}{dt} + \frac{dw}{dt}$$

would emergy in many forms, such as

$$\frac{d}{dt}(v +_v w) = \frac{dv}{dt} +_a \frac{dw}{dt} \qquad ,$$

in each of which the distribution law has practically been destroyed. Secondly, the physicist would point out that in every physical context the subscripts of the addition symbols are really redundant because they follow from the type of vectors added. And, finally, he would point out that the use of a single addition symbol never seduces him to add a velocity to a force since the incentive to do so never arises. The defence is purely pragmatic.

The physicist goes further. With respect to a point mass m , he introduces a gravitational potential G , which in some considerations is treated as a single physical object, for instance, in the sentence "the dimension of the gravitational potential is $length^2/time^2$ ". Also potentials can be "added" by the physicist: a system consisting of two point masses, more precisely of point mass $m0$ with gravitational potential $G0$ and a point mass $m1$ with gravitational potential $G1$, gives rise to a gravitational potential $G0 + G1$. What kind of addition is that?

A possible answer is to shrug one's shoulders and to say "Mostly the usual one: it is symmetric and associative and there is a zero potential 0 satisfying $G + 0 = G$ for any potential G . Furthermore, it has some special properties in connection with other operators that are specific for potentials, e.g., the nabla operator ∇ distributes over it:

$$\nabla(G0 + G1) = \nabla G0 + \nabla G1 \qquad ,$$

but that is really another story that more belongs to the nabla operator to be introduced later.".

The helpful physicist will certainly give you a much more detailed answer: he will tell you that a potential assigns a scalar value to each point of three-dimensional space and, conversely, is fully determined by those values —i.e., two potentials are the same potential if and only if they are everywhere equal— ; he will furthermore tell you that, by definition, in any point of three-dimensional space, the value of $G0 + G1$ equals the sum of the values of $G0$ and $G1$ in that point. By the convention of "point-wise addition" he thus defines the addition of potentials in terms of addition of real numbers.

Remark In the same vein he will define the nabla operator as a differentiation operator. That the nabla distributes over addition then emerges as a theorem. (*End of Remark.*)

Once we have chosen a coordinate system for the three-dimensional Euclidean space, say three orthogonal coordinates x , y , z , another view of a potential presents itself. We then have the option of viewing the potential as an expression in the coordinates, i.e., we equate for some function g ,

$$G = g.(x,y,z) \qquad ,$$

where the function is such that, for any triple (a,b,c) , the value of $g.(a,b,c)$ equals the value of the potential G at the point that has that triple (a,b,c) as its coordinates.

This last view gives a very familiar interpretation to the plus sign in $G0 + G1$: the latter formula now stands for the expression

$$g0.(x,y,z) + g1.(x,y,z) \qquad ,$$

i.e., our plus sign just adds two expressions to form a new expression that is their sum.

This view thus allows a very familiar interpretation for the plus sign in $G0 + G1$; the price we have to pay for that convenience is the introduction of names, like $G0$ and $G1$, that stand for expressions in, say, x , y , and z but, being just names, do not state that dependence explicitly (as a functional notation like $g.(x,y,z)$ would have done).

Modern mathematical usage freely introduces names for all sorts of mathematical objects such as sets, points, lines, functions, relations, and alphabets, but is reluctant to introduce a name for an expression in a number of variables. There is a very good reason for that reluctance: because of their hidden dependence on some variables, those names may become quite tricky to manipulate. (An example of how tricky it may become is provided by the names "yesterday", "today", and "tomorrow", which admit sentences such as "Tomorrow, today will be yesterday.".)

When physicists call a potential "G" , they do precisely what, for good reasons, the mathematicians are very reluctant to do; they adopt a mathematically dubious convention. The reason why they get away with it —at least most of the time— is probably that three-dimensional space (plus time) is the standard context in which almost all of classical physics is to be understood. With that understanding, any notation —like $g.(x,y,z)$ — that indicates that dependence explicitly is unnecessarily lengthy.

We mentioned the physicists because we are partly in the same position as they are. For the sake of coping with programming language semantics we shall develop a theory, and we may seem to have adopted the "dubious convention" of the physicists in the sense that, when the theory is applied to programming language semantics, things that were denoted by a name in our theory stand in the application for expressions in programming variables.

Remark Once we have chosen a set of Cartesian coordinates for the three-dimensional space, we have introduced a one-to-one correspondence between the points of space and the triples of coordinate values. For programs, this has given rise to the metaphor that is known as "the state space". For a program that operates on n variables, the corresponding "state space" is visualized as an n-dimensional space with the n variables of the program as n Cartesian coordinates. Thus the metaphor introduces a one-to-one correspondence between the points in state space and all combinations of values for the n variables. Since each combination of these values corresponds to a state of the store consisting of those variables, there is a one-to-one correspondence between the states of the store and the points in state space; hence the latter's name. It is a well-established metaphor, and we shall use it freely. Instead of "for any combination of values of the program variables" we often say "for any point in state space", or "everywhere in state space" or "everywhere" for short. Another benefit of the metaphor is that we can describe the sequence of states corresponding to a computation as a "path" through state space, which is traversed by that computation. Furthermore, the metaphor allows us to view certain program transformations (in which program variables are replaced by others) as coordinate transformations.

In short, the notion of state space has its use. The reader that encounters the term for the first time should realize that this use of the term "space" represents a considerable generalization of normal three-dimensional space: there is in general nothing three-dimensional about a state space, its coordinates are rarely real, and Euclidean distance between two states is not a meaningful concept (not even if the program variables are of type integer and the state space can be viewed as consisting of grid points). (*End of Remark.*)

Upon closer inspection it will transpire that our convention is not half as dubious as it may appear at first sight. Hidden dependence on a variable makes manipulation tricky only in contexts in which the variable occurs explicitly as well. We are safe in the sense that our theory is developed independently of its application, and that it is only in the application that names occurring in the theory are made to stand for expressions in program variables. We beg the reader to remember that the state space is not an intrinsic ingredient of our theory and that it only enters the picture when we apply our theory to programming language semantics.

In fact we go further and urge the reader to try to read the formulae of our theory without interpreting names as expressions in the coordinates of some state space. Such an interpretation is not only not helpful, because it is confusing, but is even dangerous, because expressions on a state space provide an overspecific model for what our theory is about, and as a result the interpretation might inadvertently import relations that hold for the specific model but do not belong to the theory.

Because the theory is to be understood independently of its application to programming language semantics it would not do to call those names in the theory variables of type "expression". We need another, less committing, term. After quite a few experiments and considerable hesitation, we have decided to call them variables of type "structure". So, in our theory we shall use names to stand for "structures".

Our notion of a "structure" is an abstraction of expressions in program variables in the sense that the state space with its individually named dimensions has been eliminated from the picture. We do retain, however, that expressions in program variables have types, i.e., they are boolean expressions or integer expressions, etc.; similarly our theory will distinguish between "boolean structures", "integer structures", etc.

The reader who is beginning to wonder what our structures "really" are should control his impatience. The proper answer to the question consists in the rules of manipulation of formulae containing variables of type structure. In due time, these rules will be given in full for boolean structures, which are by far the most important structures with which we shall deal. (We leave to the reader the exercise of moulding Peano's Axioms into the manipulative rules for integer structures.) The impatient reader should bear in mind that this introductory chapter's main purpose is to give the reader some feeling for our goals and to evoke some sympathy for our notational decisions.

$$*   *   *$$

The next notational hurdle to take has to do with the notion of equality. Difficulty with the notion of equality might surprise the unsuspecting reader, who feels —not without justification— that equality is one of the most fundamental, one of the most "natural", relations. But that is precisely the source of the trouble! The notion of equality came so "naturally" that for many centuries it was quite common not to express it explicitly at all.

For instance, in Latin, in which the verb "esse" for "to be" exists, it is not unusual to omit it (e.g., "Homo homini lupus."). In mathematical contexts equality has been expressed for ages either verbally or implicitly, e.g., by writing two expressions on the same line and leaving the conclusion to the intelligent reader. We had in fact to wait until 1557, when Robert Recorde introduced in his "Whetstone of witte" the —in shape consciously designed!— symbol $=$ to denote equality. In the words of E. T. Bell: "It remained for Recorde to do the right thing.".

Yes, Recorde did the right thing, but it took some time before it began to sink in. It took in fact another three centuries before the equality sign gained in principle the full status of an infix operator that assigned a value to expressions of the form $a = b$. The landmark was the publication of George Boole's "Laws of Thought" in 1854 (fully titled *An Investigation of The Laws of Thought on which are founded The Mathematical Theories of Logic and Probabilities*). Here, Boole introduced what is now known as "the boolean domain", a domain comprising two values commonly denoted by "*true*" and "*false*" , respectively. In doing so, he gave $a = b$ the status of what we now call a "boolean expression".

Yes, Boole too did the right thing, but we should not be amazed that his invention, being less than 150 years old, has not sunk in yet and that, by and large, the boolean values are still treated as second-class citizens. (We should not be amazed at all, for the rate at which mankind can absorb progress is strictly limited. Remember that Europe's conversion from Roman numerals to the decimal notation of Hindu arithmetic took at least six centuries.)

After this historical detour, let us return for a moment to our physicist with his potentials. In terms of two given potentials $G0$ and $G1$, he is perfectly willing to define a new potential given by

$$G = G0 + G1 \qquad .$$

Similarly, a mathematician is perfectly willing to define in terms of two given 10-by-10 real matrices A and B a new 10-by-10 real matrix C by

$$C = A + B \qquad .$$

We said "similarly", and rightly so. A potential associates a real value with each of the points of three-dimensional, "physical" space; a 10-by-10 real matrix associates a real value with each of the 100 points of the two-dimensional 10-by-10 space spanned by the row index and the column index. The additions of potentials and of matrices are "similar" in the sense that they are to be understood as "point-wise" additions. That in the case of matrices, where the underlying space is discrete, it is customary to talk about "element-wise" addition is in this case an irrelevant linguistic distinction.

With this convention of point-wise application in mind it would have stood to reason to let $A = B$ stand for the 10-by-10 boolean matrix formed by element-wise comparison.

But this is not what happened. People were at the time unfamiliar with the boolean domain. Consequently, a boolean matrix was beyond their vision and hence a formula like the above $C = A + B$ was not read as a boolean matrix but was without hesitation read as a statement of fact, viz., the fact that C and $A + B$ were "everywhere" —i.e., element by element— equal. Similarly, $G = G0 + G1$ is interpreted as a statement of fact, viz., the fact that the potentials G and $G0 + G1$ are "everywhere" equal and not as an expression for a "boolean potential", *true* wherever G equals $G0 + G1$ and *false* elsewhere.

For real expressions A and B , $A = B$ stands for a boolean expression and, hence, for real structures we would like $A = B$ to denote a boolean structure, but the conventional interpretation reads $A = B$ as the fact that A and B are "everywhere" equal, that A and B are "the same". In retrospect we can consider this conventional interpretation an anomaly, but that is not the point. The point is: can we live with that anomaly in the same way as physicists have lived with it since potentials were invented and mathematicians have lived with it since matrices were invented?

It would be convenient if the answer were affirmative, as might be the case in a context in which there is no need for dealing with boolean structures. But ours is a very different context, for almost all our structures will be boolean. Hence we cannot live with the notational anomaly and we have to do something about it.

One could try to find a way out by introducing two different equality signs, one to express that two structures are "the same" and the other to denote the boolean structure formed by the analogue of point-wise comparison. This, however, would lead to an explosion of symbols because the same dilemma presents itself in the interpretation of expressions like $A < B$, $A \leq B$, etc.

For the description of our way out of the dilemma we first introduce a little bit of terminology. For integers x and y we all recognize $x + y$ and $x - y$ as integer expressions and $x < y$ and $x = y$ as boolean expressions; they are expressions in x and y. Note that 7 and 5 are also integer expressions and that *true* and *false* are boolean ones. The latter ones distinguish themselves from the former ones by the circumstance that they are expressions in no variables at all. We distinguish the integer constants from the general integer structures by calling them "integer scalars". Similarly, we distinguish the boolean constants *true* and *false* from the general boolean structures by calling them "boolean scalars". A space of zero dimensions is called "the trivial space". (It is not called "the empty space" because, just as the empty product —i.e., the product of zero factors— is most conveniently defined to be equal to 1, the space of zero dimensions is most conveniently considered to consist of a single anonymous point. Since there is nothing to distinguish two anonymous points from each other, the trivial space is unique.) When convenient, we describe our "scalars" also as "structures on the trivial space".

After these preliminaries, we can describe our way out of the dilemma.

Firstly, two structures connected by a relational operator that can be applied element-wise —such as $A = B$ or $X \Rightarrow Y$— is, in deviation from tradition, not interpreted as statement of fact but denotes a boolean structure, in principle as general as the connected operands.

Secondly, we introduce one function from boolean structures to boolean scalars. It is called the "everywhere" operator and its application is denoted by placing a pair of square brackets around the boolean structure to which it is applied. Applied to a boolean scalar, the "everywhere" operator acts as the identity function; because *true* and *false* are the only boolean scalars, this can be summarized by

$$[true] = true \qquad \text{and} \qquad [false] = false \qquad .$$

Treating potentials as structures, we would replace the physicist's definition of G by

$$[G = G0 + G1] \qquad .$$

Similarly, treating matrices as structures, we would replace the mathematician's definition of C by

$$[C = A + B] \qquad .$$

The main price we have to pay for our convention is a pair of additional square brackets whenever we have to express "complete" equality between two operands that might be structures. In passing, we remark that we shall use essentially only a single property of equality, viz., that it is preserved under function application. This principle is known as the Rule of Leibniz, and is traditionally written as

$$x = y \;\Rightarrow\; f(x) = f(y) \qquad .$$

We shall write the Rule of Leibniz in the following fashion:

$$[x = y] \Rightarrow [f.x = f.y] \qquad .$$

The first pair of square brackets is needed because the arguments x and y could be structures; the second pair of square brackets is needed because f might be structure valued. Finally, we have adopted in our rewrite the convention of indicating functional application by an infix full stop (a period).

Remark As said, the single property of equality that will be used is that it is preserved under function application. Conversely, the one and only property of function application that will be used is that it preserves equality. We shall return to this later. Here it suffices to point out that, evidently, function application and equality are so intimately tied together that the one is useless without the other. Our decision to introduce an explicit symbol for functional application is the natural counterpart of Recorde's decision to do so for equality. (*End of Remark.*)

One more notational remark about equality. In the case of boolean operands, we admit besides $=$ an alternative symbol for the equality, viz., $\equiv$. We do so for more than historical reasons.

The minor reason for doing so is purely opportunistic, as it gives us a way out of the traditional syntactic dilemma of which binding power to assign to the infix equality operator. By assigning to $=$ a higher binding power than to the logical connectives, and to $\equiv$ the lowest binding power of all logical connectives, we enable ourselves to write for instance

$$[x = y \;\equiv\; x \leq y \,\wedge\, x \geq y]$$

without any further parentheses.

The major reason is much more fundamental. In the case of boolean operands, equality —in this case also called "equivalence"— enjoys a property not shared by equality in general: equivalence is associative. The importance of a property like associativity justifies both a special name and a special symbol for equality between boolean operands. Therefore we shall use $\equiv$ for the equality between boolean operands; if the operands might be boolean but need not be, we shall use Robert Recorde's $=$.

*   *   *

With A , B , C , and D four real structures, (A,B) is a pair of real structures, and so are (C,D) and $(A + C , B + D)$. We wish to consider the last one the sum of the first two, in formula

$$[(A,B) + (C,D) = (A + C , B + D)] \qquad ,$$

i.e., addition —and in general: each operator that admits point-wise application— distributes over pair-forming (in general: over tuple-forming). This also holds for equality itself, i.e., instead of

$$[(A,B) = (C,D)]$$

we may write equivalently

$$[(A = C , B = D)] \qquad ,$$

an expression of the form

$$[(X,Y)]$$

for X and Y of type boolean structure. So far, we applied the "everywhere" operator only to a boolean structure. What does it mean to apply the "everywhere" operator to a pair? By definition it means

$$[(X,Y)] \equiv [X] \wedge [Y] \qquad .$$

With the "everywhere" operator read as universal quantification, this relation reflects that we can read the expression (X,Y) in two ways: we can read it as a pair of boolean structures, and we can read it as a single boolean structure on a "doubled" space, viz., the original space with an added two-valued dimension (whose two coordinate values could be "left" and "right"). The $\wedge$ in the last formula stands for the universal quantification over that added dimension.

Equating n-tuple forming of structures with the formation of one structure, but over a space with an added n-valued dimension, will give rise to noticeable simplifications, both conceptual and notational.

CHAPTER 2

On substitution and replacement

With respect to substitution and replacement, two extreme attitudes seem to prevail. Either the manipulations are deemed so simple that the author performs them without any explanation or statement of the rules, or the author tries to give a precise statement of the most general rules and ends up with 200 pages of small print, in which all simplicity has been lost. Neither is entirely satisfactory for the purposes of this little monograph.

The first extreme has to be rejected in view of our stated goal of presenting fairly calculational arguments. The "mindless" manipulation of uninterpreted formulae that is implied by "letting the symbols do the work" is clearly impossible without a sufficiently clear view of the permissible manipulations.

The second extreme has to be rejected because this is not a treatise on formal logic and because it would probably provide much more than needed for the limited scope of the theory we have set out to develop.

Rightly or wrongly, we assume that the reader can do without a lot of the formal detail of the lowest level. For instance, while we firmly intend to state explicitly the relative binding powers of all operators so as to make our formulae sufficiently —i.e., semantically— unambiguous, we shall not give a formal definition of the grammar. Also, though the reader is supposed to be able to parse our formulae, we shall not give him an explicit parsing algorithm for them. To simplify the parsing job, we have adopted a context-free grammar in which the scope of dummies is delineated by an explicit parenthesis pair. Parsing will furthermore be aided by layout. And, as said, rightly or wrongly we assume that these simple measures will suffice.

In the remainder of this chapter we shall try to sail between Scylla and Charybdis by confining ourselves mainly to the identification of the sources of complication. Awareness of them should suffice to protect the human formula manipulator against making the "silly" mistakes.

$$*\quad*\quad*$$

It all starts innocently enough. Given

(0) $13 = 5 + 8$ and

(1) $x > 13$,

we all feel pretty confident in concluding

(2) $x > 5 + 8$;

the equality (0) tells us that in (1) we may replace 13 by $5 + 8$ without changing (1)'s value. We are, however, courting trouble as soon as we view the transition from (1) to (2) purely as a matter of string manipulation. Given

(3) $y < 131$ and

(4) $z = q - 13$,

we should conclude neither

(5) $y < 5 + 81$

from (0) and (3) nor

(6) $z = q - 5 + 8$

from (0) and (4). The moral of the story is that there is more to it than just the manipulation of strings of characters.

Though strings of characters are what we write down and see, the only strings we are interested in are formulae that satisfy a formal grammar. The rôle of this grammar goes beyond defining which strings represent well-formed formulae; it also defines how a well-formed formula may be parsed. Indicating permissible parsings with parentheses, we would get, rewriting (2), (5), and (6),

(7) $x > (5 + 8)$,

(8) $y < 5 + (81)$,

(9) $z = (q - 5) + 8$.

In (7), $5 + 8$ still occurs as a subexpression; in (8) and (9) it no longer does. Relation (0) does not express equality of strings —"13" and "5 + 8" are different strings!— but equality of two (integer) expressions. In appealing to it when "replacing equals by equals" we may only replace the (integer) subexpression 13 by the (integer) subexpression $5 + 8$ or vice versa. Each expression is a function of its subexpressions, but that statement only makes sense provided that the grammar describes how each well-formed formula may be parsed, thereby identifying its subexpressions (and their types). In replacing equals by equals, the moral of the story is that, apart from the internal structures of the subexpressions exchanged, the two expressions equated should admit the same parsings. The consequences of this require-ment obviously depend on the grammar adopted: (6) would have been a valid conclusion for a grammar in which $+$ has been given a higher binding power than $-$, whereas in a grammar in which $+$ and $-$ are left-associative, the proper conclusion would have been

$$z = q - (5 + 8)$$.

Remark We don't require the grammar to be unambiguous. For instance, the associativity of the multiplication can be dealt with syntactically by means of a production rule like

$$\langle product \rangle ::= \langle factor \rangle$$

$$| \langle product \rangle \cdot \langle product \rangle$$.

We feel reasonably confident that, with a suitable adaptation of the notion of a formal grammar, also the symmetry of the multiplication can be dealt with syntactically. (*End of Remark*.)

The important thing to remember is that, while the manipulating mathe-matician sees and writes only strings, he actually manipulates the parsed formulae. Consequently, a grammar that leads to a complicated parsing algorithm adds to the burden of manipulation. We have therefore adopted a systematic and relatively simple grammar that admits context-free parsing.

At this stage the reader may wonder why we have adhered to the familiar infix notation, together with the traditional mechanism of relative binding powers so as to reduce the need of parentheses. To tell the truth: sometimes, so do we. We certainly appreciate the smoothness with which infix notation caters for rendering associativity, viz., omission of parentheses. Evidently, we felt the burden of binding powers too light to justify a deviation from the familiar.

*   *   *

We get another complication with grammars in which variables may have a limited scope, such as the dummy i in the quantified expression $(\forall i:: Q.i)$. Here the parenthesis pair delineates the scope of the dummy i , which is therefore said to be "local to this expression"; a variable that occurs in the text of an expression but is not local to it or to one of its subexpressions is said to be "global to that expression" or, alternatively, to be "local to that expression's environment". (For "local" and "global", also the terms "bound" and "free", respectively, are used.) The formal definition of the set of variables global to an expression is a straightforward recursive definition over the grammar.

Variables are place holders in the sense that their main purpose is to get something substituted for them. More precisely, when we substitute something for a variable, *each* occurrence of that variable is replaced by that *same* something. Identifying different variables by different identifiers is a means for indicating which places have together to be filled by the same something in the case of substitution. For instance, substitutions for the variable X in

$$[2 \cdot X = X + X]$$

may yield

$$[2 \cdot Y = Y + Y] \qquad \text{or}$$

$$[2 \cdot (Y - Z) = (Y - Z) + (Y - Z)]$$

but never

$$[2 \cdot (Y - Z) = (Y - Z) + X] \qquad ,$$

since in the latter case the replacement by $(Y - Z)$ has not been applied to all occurrences of X .

In the absence of local variables, all variables are global and life is simple. There are as many variables as identifiers and they are in one-to-one correspondence to each other to the extent that we hardly need to distinguish between the variable and the identifier identifying it.

But consider now an expression like

(10) $(\forall i:: P.i) \lor (\exists j:: Q.j)$.

This expression has four variables: P , Q , i , and j . The first two, P and Q , are global to expression (10), i.e., this expression is deemed to occur in an environment that may have more occurrences of P and Q , and substitution of something for P (or Q) should be applied to *all* occurrences of P (or Q), both in (10) and in its environment. As long as the environment of an expression like (10) is left open, for each variable global to that expression the set of its occurrences is open-ended.

The other two variables, i and j , are sealed off, so to speak: their sets of occurrences are closed. The first one, introduced after the universal quantifier $\forall$, is local to the first disjunct, and the second one, introduced after the existential quantifier $\exists$, is local to the second disjunct. Their occurrences are by definition confined to their respective scopes as delineated by the parenthesis pairs.

In the environment of (10), P and Q play different rôles: P refers to the term in the universally quantified expression and Q to the term in the existentially quantified expression. In contrast, the variables identified in (10) by i and j play in the environment no rôle at all. As a result, it should not be very relevant for the environment which identifiers have been used for these local variables. In fact, instead of (10) we could equally well have written

(11) $(\forall i:: \ P.i) \vee (\exists k:: \ Q.k)$

or even

(12) $(\forall i:: \ P.i) \vee (\exists i:: \ Q.i)$.

Since a variable local to an expression plays by definition no rôle in that expression's environment, we can always choose a fresh identifier to denote it. In (12) we have carried that freedom to its ultimate consequence and have used the same identifier for both local variables. No confusion arises, since it is still quite clear over which variable the universal quantification takes place and over which variable the existential quantification.

Note that we have taken a major step. The one-to-one correspondence is no longer between variable and identifier, but between variable and identifier together with its scope.

It is customary to go one step further. It so happens that (10) is equivalent to

(13) $(\forall i:: \ P.i \vee (\exists j:: \ Q.j))$.

In view of (12) we would like to permit

(14) $(\forall i:: \ P.i \vee (\exists i:: \ Q.i))$

as well. In fact we do.

Scopes, being delineated by matching parenthesis pairs, are either disjoint, as in (10), (11), and (12), or nested, as in (13) and (14). For variables that have nested scopes and that are identified by the same identifier, the additional rule is that all occurrences of that identifier within the inner scope refer to the variable with the inner scope. In the case of such multiple use of identifiers, the basic rule is that each derivation should also have been possible with each variable identified by a fresh, distinct identifier.

For instance, the equivalence of (10) and (13) is an instantiation of the axiom

(15) $[(\forall i:: \ P.i) \vee R \equiv (\forall i:: \ P.i \vee R)]$,

viz., with $(\exists j:: \ Q.j)$ substituted for R .

Asked to derive an equivalent expression for

(16) $(\forall i :: \; P.i) \; \vee \; Q.i$

—an expression in which Q is applied to a global variable identified by i — we should not fall into the trap of writing down the erroneous

$$(\forall i :: \; P.i \; \vee \; Q.i)   ,$$

in which P and Q are "accidentally" applied to the same variable. That it is an accident we see when we remember that we are free in our choice of dummies and that (15) could as well have been written

$$[(\forall i :: \; P.i) \; \vee \; R \; \equiv \; (\forall k :: \; P.k \; \vee \; R)]   .$$

Substitution in the latter of $Q.i$ for R would yield the correct alternative for (16):

$$(\forall k :: \; P.k \; \vee \; Q.i)$$

in which Q is still applied to the global variable identified by i .

In the presentation of an axiom like (15) we feel free to omit the standard caveat "provided i is not global to R ": we should remember that the dummy of the right-hand quantification is to be understood as a fresh identifier. We prefer to consider the caveat as "obvious".

Remark That last desire is perhaps very naive in view of the amount of effort logicians have spent on explaining "bound" versus "free" variables. They consider perhaps more complicated manipulations than we shall do. From the moment that we decided to delineate the scopes of local variables explicitly by an obligatory parenthesis pair and to adopt a syntax that unambiguously identifies the local variables of each scope —and this decision was taken many years ago— we never made an error due to clashing identifiers. One manipulates under the constant awareness that within its scope each variable could have been identified by a fresh identifier. It is a permanent supervision, not unlike the physicist's constant awareness that his formulae should be invariant under a change of units or of coordinate system: renaming of variables is a similarly irrelevant transformation. (*End of Remark.*)

CHAPTER 3

On functions and equality

We have to say a word or two about functions because, depending on the ways permitted for their definition, it may be questionable whether they are well-defined at all. Fortunately, our functions will be simple: they will be total functions with values and arguments of well-defined types.

The restriction to total functions is completely in line with the whole idea of defining programming language semantics by means of weakest preconditions. For deterministic programming languages, one can consider the final machine state as a function of the initial state, and one may try to define the programming language semantics by defining how a program determines that function. One then encounters the problem that in general that function is a partial function of the initial state: for some initial states, the program fails to terminate and, consequently, the corresponding final state is to be regarded as undefined. As we shall see later, weakest preconditions —which embody a form of "backwards reasoning"— circumvent that problem by being total functions of the postcondition.

The arguments of the functions are of familiar types, such as boolean or integer scalars —which we deem well-understood— and boolean or integer structures. The function values are of equally familiar types. Each new function we introduce is a total mapping from a previously introduced domain to a previously introduced domain.

Aside An integer function that accepts itself as argument, is of a type α that would satisfy the relation

$$\alpha = (\alpha \rightarrow int) \qquad .$$

With such a recursively defined type, it is no longer obvious whether a function definition makes sense or can lead to a contradiction. Such function definitions, fortunately, do not occur in this monograph. (*End of Aside.*)

Furthermore, our ways of defining functions are very simple. Firstly, we introduce a modest number of operators with which we can construct expressions; these expressions are postulated to be functions of their subexpressions. Furthermore, we shall define functions as solutions of an equation that contains the argument(s) of the function being defined as parameter(s). In that case we have the obligation to prove that the equation in question has a unique solution for any value of the parameter(s).

For instance, let multiplication, denoted by an infix $\cdot$, be a defined operator on integers. With integer structures x and y ,

$$x \cdot x$$

is an integer structure,

$$y = x \cdot x$$

is a boolean structure, and

$$(0) \qquad [y = x \cdot x]$$

is a boolean scalar, of which x and y are the global variables. Expression (0) is almost an equation but not yet, because it fails to identify the unknown. Our notational convention is that we can turn the boolean scalar (0) into "an equation in y" by prefixing it with "y:" . Thus

$$(1) \qquad y: [y = x \cdot x]$$

is an equation of which x is a global variable. We can now define the function "*square*" —from integer structures to integer structures of the same shape— by calling the solution of (1) "*square.x*" . Since

$$[u = x \cdot x] \wedge [v = x \cdot x] \Rightarrow [u = v] \qquad ,$$

the solution of (1) is unique, and, since $\cdot$ is a total operator on integer structures, the solution always exists.

For an equation as simple as (1), the above is a bit pompous: an equation of the form

$$y: [y = \text{"total expression not containing } y\text{"}]$$

has a unique solution, which is a total function of the global variables of the equation. In the future we shall not repeat this argument. In these simple circumstances we shall even omit the equation and say something like: "Let for integer structures x the function *square* be defined by

$$[square.x = x \cdot x] \qquad \text{."}.$$

For equations with more complicated occurrences of the unknown, uniqueness and existence of the solution will be demonstrated explicitly.

$$* \qquad * \qquad *$$

Our use of the notion of a function will be that, for f a function, we appeal to

(2) $[x = y] \Rightarrow [f.x = f.y]$,

where the left-hand "everywhere" has been introduced to cater also for the case that f's argument is a structure, and the right-hand "everywhere" because f's value might be a structure. In hints, the name "Leibniz" will indicate an appeal to rule (2).

Remark The original statement of Leibniz (who was probably aware of the existence of the identity function) seems to go further; in our notation it would be rendered by

(3) $[x = y] \equiv (\forall f:: [f.x = f.y])$.

Note that, if we render equality of the functions g and h by $[g = h]$, this is usually defined by

(4) $[g = h] \equiv (\forall x:: [g.x = h.x])$.

The pleasing symmetry between functions and arguments as displayed by (3) versus (4) will not be pursued. (*End of Remark.*)

There is another way of reading Leibniz's Rule (2): function application preserves equality.

Remark The above use of the verb "to preserve" is usually reserved for the case of ordering relations: for a function f from real structures to real structures, to be "monotonic" would mean

(5) $[x \leqslant y] \Rightarrow [f.x \leqslant f.y]$,

i.e., application of f "preserves" the ordering relation $\leqslant$. (For analogy's sake, we ignore that the notion of monotonicity has been extended to cover the situations in which the ordering relations between the arguments and between the function values are different.) Note that the analogy goes further: were we to define an ordering relation on functions of the same type as f , we would define —in analogy to (4)—

$$[g \leqslant h] \equiv (\forall x:: [g.x \leqslant h.x])   .$$

(*End of Remark.*)

We mention this because Leibniz's Rule captures not only our use of the notion of "function" but also our use of the notion of equality. The two go inextricably together: function application is what preserves equality, but also, equality is what is preserved by function application. Hence, also, the title of this chapter.

We must finally draw attention to a distinction we have to make due to our introduction of structures. Let f be a function from structures to structures of the same shape. It then may be the case that, besides Leibniz's Rule (2), we have the stronger

(6) $[x = y \Rightarrow f.x = f.y]$.

(We shall see later that (6) is indeed stronger than (2), i.e., that (2) follows from (6).) Functions for which (6) holds are called "punctual functions". In terms of the model we presented in Chap. 1, the punctual functions correspond to operators that are applied element-wise; taking the negative of a matrix —i.e., changing the sign of each element— is a punctual function, whereas taking the transpose of a (square) matrix —i.e., interchanging rows and columns— is *not* a punctual function.

In hints appealing to (6) we shall draw attention to the punctuality of the function in question.

CHAPTER 4

On our proof format

This chapter describes how we present our calculational proofs and, to a certain extent, why we have chosen to adopt this format.

The statement of a theorem is in essence a boolean expression, and its proof is in essence a calculation evaluating that boolean expression to the value *true* . This is the sense in which the vast majority of our proofs are calculational. The most straightforward way of evaluating a boolean expression is to subject it to one or more value-preserving transformations until *true* or *false* is reached. The reason that we may need a number of consecutive transformations is that we wish to confine them to manipulations from a restricted repertoire.

Remark We will not restrict that repertoire of manipulations to the absolute minimum because that would not be practical for our purposes: it would make our calculations much longer than we would like them to be. It would be like doing mathematics without theorems, reducing each argument down to the axioms, and that is not practical. The reader should bear in mind that our purposes —i.e., using logic— are quite different from those of the logicians that study logics: for their purposes it might be appropriate to reduce logical systems to their bare essentials, but for our purposes it is not. (*End of Remark.*)

Let [*A*] denote the boolean expression to be evaluated, and let the evaluation take, say, 3 steps, which would amount to the occurrence of 2 intermediate results. More precisely, for some expressions $[B]$ and $[C]$:

the first step would establish $[A] \equiv [B]$;
the second step would establish $[B] \equiv [C]$;
the third step would establish $[C] \equiv true$;

thus the whole calculation would establish $[A] \equiv true$. (Had the third step established $[C] \equiv false$, the whole calculation would have established $[A] \equiv false$.)

In the above rendering of the calculation, the intermediate expressions $[B]$ and $[C]$ each occur twice. Since, in general, B and C can be quite elaborate expressions, we need for brevity's sake a format in which such repetitions are avoided. We propose

$[A]$
$=$ {hint why $[A] \equiv [B]$ }
$[B]$
$=$ {hint why $[B] \equiv [C]$ }
$[C]$
$=$ {hint why $[C] \equiv true$ }
$true$.

In the above, the actual hints between the braces would not mention the expressions A , B , or C ; they would identify the value-preserving transformation.

Remark The virtue of this format is more than brevity alone, for it allows us to conclude $[A] \equiv true$ without reading the intermediate expressions. In our former rendering one would have to check that the right-hand side of $[A] \equiv [B]$ is the same expression as the left-hand side of $[B] \equiv [C]$. Our proof format expresses that sameness by syntactic means. (*End of Remark*.)

Had we no further interest in brevity, the above format would suffice. But we do have an interest in further brevity, and as a result our format can do with some refinements, as the following example shows. Let our proof obligation be of the form $[A = D]$. Rendered in the above format, the calculation could be of the form

$[A = D]$
$=$ {hint why $[A = B]$ }
$[B = D]$
$=$ {hint why $[B = C]$ }
$[C = D]$
$=$ {hint why $[C = D]$ }
$true$.

Note Had we strictly followed the previous example, the first hint would have been "hint why $[A = D] \equiv [B = D]$ ", and similarly for the second hint. Thanks to Leibniz's Rule, the hints given here suffice to justify the first two transformations as value-preserving. Since $[C = D]$ is the same as $[C = D] \equiv true$, the third hint exactly follows the previous example. (*End of Note.*)

The above is not very nice if the expression here denoted by D is lengthy: it has to be repeated from line to line. We therefore propose to render this calculation in the form

$$A$$
$$= \{\text{hint why} [A = B] \}$$
$$B$$
$$= \{\text{hint why} [B = C] \}$$
$$C$$
$$= \{\text{hint why} [C = D] \}$$
$$D   ,$$

in which the successive steps are no longer confined to value-preserving transformations from one scalar to the next, but may be between structures on some space. The complementary convention is that the "everywhere" operator should be applied to the relational operators to the left: the above calculation amounts to the assertion

$$[A = B] \wedge [B = C] \wedge [C = D]$$

and justifies, again without reading the intermediate results B and C , the conclusion $[A = D]$, square brackets included. Since "everywhere" has no effect —i.e., is the identity operator— on boolean scalars, this convention of the additional square brackets does not invalidate our first proof, where we had to prove $[A] \equiv true$.

Remark Because, as we shall see later, the "everywhere" operator distributes over conjunction, we could also have written the above summarizing assertion as

$$[A = B \wedge B = C \wedge C = D]   .$$

(*End of Remark.*)

So much for the fact that the steps of our calculations need not transform scalars but may transform structures as well. The structures occurring in our theory are almost exclusively boolean structures. The remaining notational refinements to be introduced in this chapter concern that special case.

In proofs of $[A \equiv D]$ by means of such a "continued equivalence" as illustrated above, it is not uncommon that $A \lor D$ occurs as intermediate expression. In our format we would get something like

$$A$$
$=$ {hint}
$$\ldots$$
$=$ {hint}
$$A \lor D$$
$=$ {hint}
$$\ldots$$
$=$ {hint}
$$D \qquad ,$$

asserting $[A \equiv D]$ on account of

$$[A \equiv A \lor D] \land [A \lor D \equiv D] \qquad .$$

Since the first conjunct is the same as $[D \Rightarrow A]$ and the second one is the same as $[A \Rightarrow D]$, such a proof of equivalence boils down to a proof "by mutual implication" —informally also known as "a ping-pong argument"— . We often prove the two implications separately.

Remark Many texts on theorem proving seem to suggest that a ping-pong argument is —if not the only way— the preferred way of demonstrating equivalence. This opinion is not confirmed by our experience: avoiding unnecessary ping-pong arguments has turned out to be a powerful way of shortening proofs. At this stage the methodological advice is to avoid the ping-pong argument if you can, but to learn to recognize the situations in which it is appropriate. (*End of Remark.*)

There are two reasons why for the genuine ping-pong argument the above format is inadequate. The first one is revealed by considering a proof of $[A \equiv D]$ in which $A \land D$ is one of the intermediate expressions:

$$A$$
$=$ {hint}
$$\ldots$$
$=$ {hint}
$$A \land D$$
$=$ {hint}
$$\ldots$$
$=$ {hint}
$$D \qquad ,$$

which asserts $[A \equiv D]$ on account of

$$[A \equiv A \land D] \land [A \land D \equiv D] \qquad .$$

The first conjunct, however, is the same as $[A \Rightarrow D]$ and the second one is the same as $[D \Rightarrow A]$, so here we have again a proof by mutual implication, but in a rather different representation. It is not so nice to have to choose between the two representations, it is nicer to have a more neutral one in between.

The second reason to refine our format is that the last two examples tend to lead to elaborate expressions with a lot of repetition between the successive lines: in the top half, A is repeated all the time, and in the bottom half D . Since a main purpose of our format is to avoid repetition, we had better do something about it. We can do so by admitting in the left column the implication sign as well.

As we shall see later, we have for any boolean structures X , Y , Z on account of Leibniz's Rule

$$[X \equiv Y] \wedge [Y \Rightarrow Z] \Rightarrow [X \Rightarrow Z] \qquad , \text{and}$$
$$[X \Rightarrow Y] \wedge [Y \equiv Z] \Rightarrow [X \Rightarrow Z] \qquad ,$$

and, on account of the transitivity of the implication,

$$[X \Rightarrow Y] \wedge [Y \Rightarrow Z] \Rightarrow [X \Rightarrow Z] \qquad .$$

Consequently, a proof of $[A \Rightarrow D]$ could, for instance, have the form

$$
\begin{array}{ll}
A & \\
\Rightarrow & \{\text{hint why } [A \Rightarrow B] \} \\
B & \\
= & \{\text{hint why } [B \equiv C] \} \\
C & \\
\Rightarrow & \{\text{hint why } [C \Rightarrow D] \} \\
D & ,
\end{array}
$$

which asserts $[A \Rightarrow D]$ on account of

$$[A \Rightarrow B] \wedge [B \equiv C] \wedge [C \Rightarrow D] \qquad .$$

Admitting $\Rightarrow$ as well as $=$ in the left column may greatly reduce the amount of writing needed. So far, so good, but in order to be really pleasant to use, our format needs still three further, independent refinements.

The first one has nothing to do with the amount of writing needed: it only allows us to write and read things in a different order. Besides the implication $\Rightarrow$, we also admit the consequence $\Leftarrow$ (read: "follows from"); from a logical point of view, it makes no difference whether we write $[A \Rightarrow D]$ or $[D \Leftarrow A]$. (It is a facility analogous to the freedom of writing the same relation either as $x \leqslant y$ or as $y \geqslant x$.) Consequently, our last example could equivalently have been rendered in the form

$$
\begin{aligned}
&D \\
\Leftarrow\quad &\{\text{hint why } [D \Leftarrow C] \} \\
&C \\
=\quad &\{\text{hint why } [C \equiv B] \} \\
&B \\
\Leftarrow\quad &\{\text{hint why } [B \Leftarrow A] \} \\
&A \qquad .
\end{aligned}
$$

Remark Often the choice between implications and consequences is irrelevant, but we have encountered many calculations in which it made a great difference. Before we introduced the consequence, we had many calculations that required considerable clairvoyance to write down in the sense that the motivation for certain manipulations would become apparent several lines further down, where everything would miraculously fall into place. Upon reading them they struck us each time as if a few rabbits had been pulled out of a hat. As soon as we realized that we ourselves had designed those proofs the other way round, we decided to present them in that direction as well, and the symbol $\Leftarrow$ was introduced. Suddenly, many a manipulation was now strongly suggested by what had already been written down. We were almost shocked to see how great a difference such a trivial change in presentation could make and came to fear that the traditional predominance of the implication over the consequence in combination with our habit of reading from left to right has greatly contributed to the general mystification of mathematics. (*End of Remark.*)

The second refinement concerns demonstranda of the form

(0) $[E] \Rightarrow [A \equiv D]$

(and is mutatis mutandis equally applicable to demonstranda of the form $[E] \Rightarrow [A \Rightarrow D]$). The above demonstrandum is equivalent to

$$[[E] \wedge A \equiv [E] \wedge D]$$

and we could therefore have used our previous format for establishing equivalences. However, this would mean that each line would start with "$[E] \wedge$ ", and that repetition is unattractive, the more so since the conjunct $[E]$ is usually needed for the justification of only a single step in the whole calculation. It is much nicer to mention $[E]$ only where it is needed. With our next refinement, the calculation for the demonstrandum (0) might be presented in the form

A
$=$ {hint why $[A \equiv B]$ }
$\quad B$
$=$ {hint why $[E] \Rightarrow [B \equiv C]$ }
$\quad C$
$=$ {hint why $[C \equiv D]$ }
$\quad D$.

At first sight, some readers may have the uneasy feeling that in the above rendering of the calculation of (0), the antecedent $[E]$ has been "smuggled in" or "hidden", but we can reassure those readers. What we have done is neither deep nor fishy: we have introduced a convenient shorthand that is quite safe to use. This format is often appropriate when the scalar $[E]$ is simply the statement that some atomic symbol enjoys some property, e.g., "f is conjunctive". Formally, such a property states the validity of some manipulations; as we shall see shortly in an example, their availability can give strong heuristic guidance.

The third convention concerns the omission of universal quantifications. If our demonstrandum is in full $(\forall x:: \ [A.x \equiv D.x])$, we might write a calculation in the form

$\qquad$ We observe for any x
$A.x$
$=$ {hint why $(\forall x:: \ [A.x \equiv B.x])$ }
$\quad B.x$
$=$ {hint why $(\forall x:: \ [B.x \equiv C.x])$ }
$\quad C.x$
$=$ {hint why $(\forall x:: \ [C.x \equiv D.x])$ }
$\quad D.x$,
which asserts $(\forall x:: \ [A.x \equiv D.x])$ on account of

$$(\forall x:: \ [A.x \equiv B.x]) \wedge (\forall x:: \ [B.x \equiv C.x]) \wedge (\forall x:: \ [C.x \equiv D.x])$$

or, equivalently

$$(\forall x:: \ [(A.x \equiv B.x) \wedge (B.x \equiv C.x) \wedge (C.x \equiv D.x)])$$.

This last notational convention is in fact completely analogous to the convention of the implied "everywhere" operator, but this time for the explicitly mentioned local variables. Similarly, it is also adopted if the demonstrandum is a universally quantified implication.

By way of example of how all this works, we shall now prove "A conjunctive predicate transformer is monotonic", a theorem that appeals to the following definitions:

(i) a function from boolean structures to boolean structures is (for historical reasons) called "a predicate transformer";

(ii) $(f$ is conjunctive$)$ $\equiv$
$(\forall X,Y:: [f.(X \wedge Y) \equiv f.X \wedge f.Y])$;

(iii) $(f$ is monotonic$)$ $\equiv$
$(\forall P,Q:: [P \Rightarrow Q] \Rightarrow [f.P \Rightarrow f.Q])$.

The proof of the monotonicity might be rendered as follows.

Proof We observe for any conjunctive predicate transformer f and any predicates P , Q

$[f.P \Rightarrow f.Q]$
$=$ {predicate calculus}
$[f.P \wedge f.Q \equiv f.P]$
$=$ { f is conjunctive}
$[f.(P \wedge Q) \equiv f.P]$
$\Leftarrow$ {Leibniz}
$[P \wedge Q \equiv P]$
$=$ {predicate calculus}
$[P \Rightarrow Q]$.

(*End of Proof.*)

The above gives some idea of the size of our steps and of the degree of detail provided in the hints.

The two hints "predicate calculus" refer to the formula

$$[X \Rightarrow Y] \equiv [X \wedge Y \equiv X]   ,$$

of which the reader is supposed to know   —after the next chapter "The calculus of boolean structures"— that it holds for any predicates X , Y . These hints refer to the "manipulations from a restricted repertoire" that we mentioned at the beginning of this chapter.

The hint "f is conjunctive" is given where the reader can be supposed to know that this means that application of f distributes over conjunction. If the situation is less familiar or the substitution is more complicated, we give the instantiation explicitly: in this case, we would have referred to the second line of (ii) "with $X,Y := P,Q$ ". When a manipulation is confined to a subexpression (as, in this case, to the left-hand side of the equivalence), our hints usually do not identify that subexpression; the reader is supposed to do so himself by seeing where the two related expressions differ.

The hint "Leibniz" refers to

$$[x = y] \Rightarrow [f.x = f.y] \qquad ,$$

i.e., the fact that function application is equality preserving.

We have a slight preference for the proof as given above compared with the proof "the other way round", which would start with

$$[P \Rightarrow Q]$$
$$= \quad \{\text{predicate calculus}\}$$
$$[P \wedge Q \equiv P]$$
$$\vdots \qquad .$$

In such a first step an (admittedly small) rabbit would have been pulled out of a hat since there are many ways of rewriting $[P \Rightarrow Q]$, such as $[P \vee Q \equiv Q]$, $[\neg P \vee Q]$, $[P \equiv \neg Q \equiv \neg P \wedge Q]$, to mention just a few. These alternatives are of course rejected because we want the conjunction to enter the picture because f is conjunctive. But when we start with $[P \Rightarrow Q]$, f is not mentioned yet! Starting, as we did, with $[f.P \Rightarrow f.Q]$, that little rabbit has been removed: now the special way of rewriting the implication can be defended on the grounds that it yields the subexpression $f.P \wedge f.Q$ of which (ii) explicitly states that we can do something with it.

Experience has taught us that in general the most complicated side is the most profitable one to start with. The probable explanation of this phenomenon is that the opportunities for simplification are usually much more restricted than the possibilities to complicate things: simplification is much more opportunity-driven than "complification". The above example, however, is too simple to give a striking illustration of this phenomenon.

CHAPTER 5

The calculus of boolean structures

In this chapter we develop the calculus of boolean structures in a rather algebraic fashion. We do so for a variety of reasons. Firstly, we have to introduce the reader to the repertoire of general formulae that will be used throughout the remainder of this booklet. Secondly, by proving all formulae that have not been postulated, we give the reader the opportunity of gently familiarizing himself with our style of conducting such calculational proofs. Thirdly, we wish to present this material in a way that does justice to how we are going to use it. Since value-preserving transformations are at the heart of our calculus, so are the notions of equality and function application; hence our desire to develop this material with the equality relation in the central rôle. (It is here that our treatment radically departs from almost all introductions to formal logic: it is not uncommon to see the equality —in the form of "if and only if"— being introduced much later as a shorthand, almost as an afterthought.)

We recall Leibniz's Rule expressing that function application preserves equality:

$$(0) \qquad [x = y] \Rightarrow [f.x = f.y] \qquad ,$$

which holds for any function f and arguments x and y of appropriate types.

Legenda We recall

- that the left-hand pair of square brackets occurs because the arguments may be structures

- that the right-hand pair of square brackets occurs because the function values may be structures
- that function application is denoted by an infix point (full stop, period) that has the highest binding power of all operators occurring in this chapter.

(*End of Legenda.*)

Appeals to (0) occur in steps such as

$$[f.x = f.y]$$
$$\Leftarrow \quad \{\text{Leibniz}\}$$
$$[x = y] \qquad , \text{ or}$$

$$[f.x = f.y]$$
$$= \quad \{\text{hint why } [x = y] \ ; \text{Leibniz}\}$$
$$true \qquad , \text{ or}$$

$$f.x$$
$$= \quad \{\text{hint why } [x = y] \ ; \text{Leibniz}\}$$
$$f.y \qquad .$$

The above illustrates the appeal to Leibniz's Rule in the case of a named function. Most of our functions, however, will be anonymous because we shall manipulate expressions and all expressions we write down are *by postulate* functions of their subexpressions. In the case of such anonymous functions, i.e., when in an expression a subexpression is replaced by an equal one, it is unusual to mention Leibniz explicitly in the hint. (Otherwise we would have to mention Leibniz in almost every step.)

$$* \qquad * \qquad *$$

In the following, we shall use the capital letters from the end of the alphabet —primarily X , Y , Z— as variables of type boolean structure. For equality between boolean operands we introduce the alternative symbol $\equiv$, which from a syntactical point of view distinguishes itself from $=$ only by the fact that (for reasons of convenience) it has been given a much lower binding power than $=$. An expression of the form $X \equiv Y$ is called "an equivalence" and is read as "X equivalent Y" or "X equivales Y".

Before pursuing further algebraic properties of the equivalence, we first turn our attention to the unary "everywhere" operator, which is denoted by surrounding the argument by a pair of square brackets. The "everywhere" operator is a function from boolean structures to boolean scalars, i.e., boolean structures on the trivial space.

In terms of the "everywhere" operator, the boolean scalars can be defined as those boolean structures for which the "everywhere" operator is the identity function, i.e., as the boolean structures solving the equation

(1) $X: [[X] \equiv X]$.

Theorem The "everywhere" operator is idempotent, i.e., for any boolean structure Y

(2) $[[[Y]] \equiv [Y]]$.

Proof We observe for any boolean structure Y

$[[Y]]$
$=$ $\{ [Y]$ is a boolean scalar, hence solves (1)$\}$
$[Y]$.

                                                              (*End of Proof.*)

After this little excursion to the "everywhere" operator and the boolean scalars, we return to the equivalence. The equivalence is a function from pairs of boolean structures to boolean structures, but also, more specifically, from pairs of boolean scalars to boolean scalars: if both X and Y are boolean structures on the trivial space, so is $X \equiv Y$. Consequently, (2) can be simplified to

$$[[Y]] \equiv [Y]   .$$

The introduction of a special symbol for equality and a special name for the equality relation in the case of boolean operands is justified by the circumstance that equivalence enjoys a property not enjoyed by equality in general: equivalence is postulated to be associative, i.e., for any X , Y , Z

(3) $[(X \equiv (Y \equiv Z)) \equiv ((X \equiv Y) \equiv Z)]$.

Together with Leibniz's Rule, (3) indeed expresses associativity: it implies that wherever we have a subexpression parenthesized as at the one side of (3)'s central $\equiv$ -sign, the parentheses may be rearranged as at its other side without changing the value of the total expression.

In our manipulations we shall never refer to (3)   —or to analogous formulae expressing the associativity of other operators— because we shall never carry out these shunting operations in such needless detail. Instead, when dealing with such "continued" expressions built with an associative operator, we shall feel free to insert or remove parenthesis pairs as we see fit.

The other important property of the equivalence —remember that $\equiv$ is only an alternative for $=$ — is that it is symmetric. (Sometimes this property is not called "symmetry" but "commutativity".) Under the convention of omitting redundant parenthesis pairs, the symmetry of the equivalence is postulated by

(4) $[X \equiv Y \equiv Y \equiv X]$.

In combination with the associativity of the equivalence, (4) is a rich formula:

- parenthesized as $[(X \equiv Y) \equiv (Y \equiv X)]$, it expresses the symmetry of the equivalence;
- parenthesized as $[X \equiv (Y \equiv Y \equiv X)]$, it expresses that $Y \equiv Y$ is a left-identity element of the equivalence;
- parenthesized as $[(X \equiv Y \equiv Y) \equiv X]$, it expresses that $Y \equiv Y$ is a right-identity element of the equivalence.

For a (not necessarily symmetric) operator that has a left- and a right-identity element, the identity element is unique.

Proof Let u be a left-identity element of the infix operator $\cdot$, and let v be a right-identity element of $\cdot$. Then we observe

u

$=$ { v is a right-identity element of $\cdot$ }

$u \cdot v$

$=$ { u is a left-identity element of $\cdot$ }

v ,

hence

(5) $[u = v]$.

Furthermore, let u' be a left- and v' be a right-identity element of $\cdot$. Then we observe

u v

$=$ {(5)} $=$ {(5)}

v u

$=$ {(5) with $u := u'$ } $=$ {(5) with $v := v'$ }

u' v' ,

hence

(6) $[u = u']$ and $[v = v']$.

Conclusions (5) and (6) settle the uniqueness of the identity element of $\cdot$.

                                                        (End of Proof.)

Hence, the equivalence has a unique identity element; its customary name is "*true*", a convention properly captured by the postulate

(7) $[X \equiv true \equiv X]$.

For a moment, one could harbour the suspicion that the equivalence has as many distinct identity elements as there are spaces on which to define structures. Fortunately, this is not the case: from (4) with the substitution $Y := [Y]$, we deduce

$$[X \equiv [Y] \equiv [Y] \equiv X]$$

and see that $[Y] \equiv [Y]$ is the identity element for the "equivalence on the space of X"; but $[Y] \equiv [Y]$, the equivalence of two boolean scalars, is a boolean scalar, i.e., a boolean structure on the trivial space. In other words, the identity element of the equivalence is as unique as the zero in $0 + x$ and $0 + y$; the identity element *true* is in fact as unique as the equivalence and the "everywhere" operator themselves.

Formulae (2), (3), (4), and (7) are boolean scalars, and, *true* being the identity element of the equivalence, we could, for instance, have written instead of (7)

$$[X \equiv true \equiv X] \equiv true \qquad ;$$

for brevity's sake we prefer (7). For the same reason we don't write

$$[[X \equiv true \equiv X]] \qquad .$$

In principle we avoid such redundant identity operators; an occasional "$\equiv true$" will occur for reasons of symmetry or analogy.

The above will be extended by more postulates, which define the properties of new operators, and by expressions that state theorems. In proving the latter, we reduce these expressions to the value *true* by using the results then available, as is illustrated in the following (simple!) proof.

Theorem Equivalence is reflexive, i.e., for any boolean structure X

(8) $[X \equiv X]$.

Proof We observe for any X

  $[X \equiv X]$
$=$ {(7), parenthesized $[X \equiv (true \equiv X)]$ }
  $[X \equiv true \equiv X]$
$=$ {(7)}
   true   .

                                                (*End of Proof.*)

Remark Note how, in the first hint, we did not feel the need to identify the subexpression being replaced. (*End of Remark.*)

Finally we derive for the identity element of the equivalence

(9)   [*true*]

by calculating for some X

$\quad$ [*true*]
$=\quad$ {(7) parsed as [*true* $\equiv (X \equiv X)$] }
$\quad$ [$X \equiv X$]
$=\quad$ {(8)}
$\quad$ *true*   ,

a result that is in accordance with our earlier conclusion that *true* is a boolean scalar. So much for the equivalence and its identity element.

$$* \quad * \quad *$$

It is time to introduce our next infix operator; it is called the disjunction, it is written as " $\vee$ " , and read as "or". We give it a higher binding power than the equivalence.

The disjunction is a function from a pair of boolean structures to a boolean structure; if both operands are scalar, so is the result.

The disjunction is postulated:

● to be symmetric, i.e., for any X , Y

(10)   [$X \vee Y \equiv Y \vee X$]   ;

● to be associative, i.e., for any X , Y , Z

(11)   [$X \vee (Y \vee Z) \equiv (X \vee Y) \vee Z$]   ;

● to be idempotent, i.e., for any X

(12)   [$X \vee X \equiv X$]   ;

● to distribute over equivalence, i.e., for any X , Y , Z

(13)   [$X \vee (Y \equiv Z) \equiv X \vee Y \equiv X \vee Z$]   .

Of the above four postulates, the first three deal with the disjunction in isolation, the last one couples the disjunction and the equivalence. The first three suffice to show our next

Theorem The disjunction distributes over itself, i.e., for any X , Y , Z

(14)   [$X \vee (Y \vee Z) \equiv (X \vee Y) \vee (X \vee Z)$]   .

Proof We observe for any X , Y , Z

$(X \vee Y) \vee (X \vee Z)$
$=$ {(11), i.e., $\vee$ is associative}
$X \vee (Y \vee X) \vee Z$
$=$ {(10), i.e., $\vee$ is symmetric}
$X \vee (X \vee Y) \vee Z$
$=$ {(11), i.e., $\vee$ is associative}
$(X \vee X) \vee (Y \vee Z)$
$=$ {(12), i.e., $\vee$ is idempotent}
$X \vee (Y \vee Z)$.

(*End of Proof.*)

Remark In the above, we have written more parentheses than usual. The "auto-distribution" we demonstrated for the disjunction holds for any operator that is symmetric, associative, and idempotent. (*End of Remark.*)

With the aid of our fourth postulate (13) we can prove

Theorem The boolean scalar *true* is a zero-element of the disjunction, i.e., for any X

(15) $[X \vee true \equiv true]$.

Remark The shorter formulation $[X \vee true]$ would have disguised *true*'s property of being a zero-element. (*End of Remark.*)

Proof The need for (13), which states a connection between disjunction and equivalence, is not surprising because, so far, we know the boolean scalar *true* only as the identity-element of the equivalence. Not surprisingly, the first step of the following calculation appeals to that definition of *true* . We observe for any X , Y

$X \vee true$
$=$ {(7) with $X := Y$ }
$X \vee (Y \equiv Y)$
$=$ {(13) with $Z := Y$ }
$X \vee Y \equiv X \vee Y$
$=$ {(7) with $X := X \vee Y$ }
$true$.

(*End of Proof.*)

Remark We would like the reader to note that, in all its simplicity, the above calculation presents a striking example of opportunity-driven simplification. (*End of Remark*.)

*   *   *

It is time to introduce our next operator. The shortest total argument would result from introducing the negation first and then defining the conjunction in terms of negation and disjunction (i.e., by postulating one of the Laws of de Morgan). We will not do so, firstly, because that is a usual order and it is always nice to show an alternative, and, secondly, because we don't want to hide the fact that the conjunction can be introduced in terms of what we already have: the negation, which is something really new, is not needed for that purpose.

Let us explore what new expressions we can write down in terms of equivalence and disjunction. Because disjunction distributes over equivalence, we can confine ourselves to variables, disjunctions of variables, and (continued) equivalences thereof. Because of the idempotence of the disjunction, the disjunctions can be confined to disjunctions of different variables. Because of the identity element of the equivalence, we can confine ourselves to (continued) equivalences of different subexpressions. Taking this, plus associativity and symmetry, into account, we conclude that in two variables we can construct, besides $X \equiv Y$ and $X \vee Y$, only two new expressions, viz., of the forms

(16) $X \vee Y \equiv Y$

(17) $X \equiv Y \equiv X \vee Y$.

The next thing to investigate is whether, viewed as functions of X and Y , these expressions have nice properties. Because (16), though shorter, looks less symmetric than (17), we shall start our investigations by exploring the properties of the latter.

The function expressed by (17) is called the conjunction; it is denoted by an infix " $\wedge$ " , which is read as "and". For reasons of symmetry (which will become clear later) we give $\wedge$ the same binding power as $\vee$. In summary, the conjunction is defined by postulating for any X , Y

(18) $[X \wedge Y \equiv X \equiv Y \equiv X \vee Y]$,

a relation also known as "The Golden Rule".

The conjunction being defined by (18) in terms of equivalence and disjunction, all its properties follow from those of the latter two operators. The conjunction is a function from a pair of boolean structures to a boolean structure because equivalence and disjunction are such functions; similarly it is a boolean scalar if both its arguments are. And, indeed, we have —as hoped— our

Theorem The conjunction is symmetric, i.e., for any X , Y

(19) $[X \wedge Y \equiv Y \wedge X]$.

Proof We observe for any X , Y

    $X \wedge Y$
= $\{(18)\}$
    $X \equiv Y \equiv X \vee Y$
= $\{$associativity and symmetry of $\equiv$ $\}$
    $Y \equiv X \equiv X \vee Y$
= $\{$symmetry of $\vee$ $\}$
    $Y \equiv X \equiv Y \vee X$
= $\{(18)$ with $X , Y := Y , X$ $\}$
    $Y \wedge X$.

(*End of Proof.*)

The conjunction has other nice properties.

Theorem The conjunction is associative, i.e., for any X , Y , Z

(20) $[X \wedge (Y \wedge Z) \equiv (X \wedge Y) \wedge Z]$.

Proof We observe for any X , Y , Z

    $X \wedge (Y \wedge Z)$
= $\{$Golden Rule with $X , Y := Y , Z$ $\}$
    $X \wedge (Y \equiv Z \equiv Y \vee Z)$
= $\{$Golden Rule with $Y := (Y \equiv Z \equiv Y \vee Z)$ $\}$
    $X \equiv Y \equiv Z \equiv Y \vee Z \equiv X \vee (Y \equiv Z \equiv Y \vee Z)$
= $\{$ $\vee$ distributes over $\equiv$; associativity of $\vee$ $\}$
    $X \equiv Y \equiv Z \equiv Y \vee Z \equiv X \vee Y \equiv X \vee Z \equiv X \vee Y \vee Z$
= $\{$associativity and symmetry of $\equiv$ $\}$
    $X \equiv Y \equiv X \vee Y \equiv Z \equiv X \vee Z \equiv Y \vee Z \equiv X \vee Y \vee Z$
= $\{$ $\vee$ distributes over $\equiv$; associativity of $\vee$ $\}$
    $X \equiv Y \equiv X \vee Y \equiv Z \equiv (X \equiv Y \equiv X \vee Y) \vee Z$
= $\{$Golden Rule with $X , Y := (X \equiv Y \equiv X \vee Y) , Z$ $\}$
    $(X \equiv Y \equiv X \vee Y) \wedge Z$
= $\{$Golden Rule$\}$
    $(X \wedge Y) \wedge Z$.

(*End of Proof.*)

Remark We have carried out this proof in such detail in order to show that it depends neither on the symmetry nor on the idempotence of the disjunction. (In the absence of symmetry, one has to postulate that $\lor$ distributes both forwards and backwards over $\equiv$.) (*End of Remark.*)

Theorem The conjunction is idempotent, i.e., for any X

(21) $[X \land X \equiv X]$.

Proof We observe for any X

$\quad X \land X$
$=$ {Golden Rule}
$\quad X \equiv X \equiv X \lor X$
$=$ {identity element of $\equiv$ }
$\quad X \lor X$
$=$ {idempotence of $\lor$ }
$\quad X$.

(*End of Proof.*)

Theorem The boolean scalar *true* is the identity element of the conjunction, i.e., for any X

(22) $[X \land true \equiv X]$.

Proof We observe for any X

$\quad X \land true$
$=$ {Golden Rule}
$\quad X \equiv true \equiv X \lor true$
$=$ {(15), i.e., zero-element of $\lor$ }
$\quad X \equiv true \equiv true$
$=$ {identity element of $\equiv$ }
$\quad X$.

(*End of Proof.*)

In each of the above four proofs a property of the conjunction is derived from the same (or a similar) property of the disjunction. The following three theorems highlight the symmetry further.

Theorem Conjunction and disjunction satisfy the Laws of Absorption, i.e., for any X, Y

(23) $[X \land (X \lor Y) \equiv X]$

(24) $[X \lor (X \land Y) \equiv X]$.

Proof We observe for any X , Y

$X \wedge (X \vee Y)$
= {Golden Rule}
$X \equiv X \vee Y \equiv X \vee X \vee Y$
= {idempotence of $\vee$ }
$X \equiv X \vee Y \equiv X \vee Y$
= {identity element of $\equiv$ }
X ,

which establishes (23); by interchanging $\wedge$ and $\vee$ we obtain the proof of (24).

(End of Proof.)

Remark In the first step of the above proof, an alternative application of the Golden Rule would have been to rewrite $(X \vee Y)$, but that would have been a blind alley, since we don't know how to manipulate the equivalence as conjunct. In the above calculation this problem is circumvented by using the Golden Rule to eliminate the outer $\wedge$. See also the proof of the next theorem. *(End of Remark.)*

Theorem Disjunction distributes over conjunction, i.e., for any X , Y , Z

(25) $[X \vee (Y \wedge Z) \equiv (X \vee Y) \wedge (X \vee Z)]$.

Proof We observe for any X , Y , Z

$(X \vee Y) \wedge (X \vee Z)$
= {Golden Rule}
$X \vee Y \equiv X \vee Z \equiv X \vee Y \vee X \vee Z$
= {properties of $\vee$ }
$X \vee Y \equiv X \vee Z \equiv X \vee Y \vee Z$
= {$\vee$ distributes over $\equiv$ }
$X \vee (Y \equiv Z \equiv Y \vee Z)$
= {Golden Rule}
$X \vee (Y \wedge Z)$.

(End of Proof.)

Because, in general, conjunction does not distribute over equivalence, we cannot prove the next, dual, theorem by interchanging $\wedge$ and $\vee$ in the above calculation. Instead, it can be proved using the previous two theorems.

Theorem Conjunction distributes over disjunction, i.e., for any X , Y , Z

(26) $[X \wedge (Y \vee Z) \equiv (X \wedge Y) \vee (X \wedge Z)]$.

Proof We observe for any X , Y , Z

$(X \wedge Y) \vee (X \wedge Z)$
= {$\vee$ distributes over $\wedge$ }
$((X \wedge Y) \vee X) \wedge ((X \wedge Y) \vee Z)$
= {Law of Absorption}
$X \wedge ((X \wedge Y) \vee Z)$
= {$\vee$ distributes over $\wedge$ }
$X \wedge (X \vee Z) \wedge (Y \vee Z)$
= {Law of Absorption}
$X \wedge (Y \vee Z)$.

(End of Proof.)

Our next theorem shows how to eliminate the equivalence as conjunct.

Theorem We have for any X , Y , Z

(27) $[X \wedge (Y \equiv Z) \equiv X \wedge Y \equiv X \wedge Z \equiv X]$.

Proof We observe for any X , Y , Z

$X \wedge (Y \equiv Z)$
= {Golden Rule}
$X \equiv Y \equiv Z \equiv X \vee (Y \equiv Z)$
= { $\vee$ distributes over equivalence}
$X \equiv Y \equiv Z \equiv X \vee Y \equiv X \vee Z$
= {rearranging and grouping}
$(X \equiv Y \equiv X \vee Y) \equiv (Z \equiv X \vee Z)$
= {Golden Rule, twice}
$X \wedge Y \equiv X \wedge Z \equiv X$.

(End of Proof.)

Note how the final " $\equiv X$" in (27) destroys the distribution of $\wedge$ over $\equiv$. As a corollary of the previous theorem —by applying it twice— the reader may derive

(28) $[W \wedge (X \equiv Y \equiv Z) \equiv W \wedge X \equiv W \wedge Y \equiv W \wedge Z]$,

a relation that can be referred to by " $\wedge$ distributes over $\equiv \equiv$ ".

Finally, we demonstrate the

Theorem For any X , Y

(29) $[X \wedge (X \equiv Y) \equiv X \wedge Y]$.

Proof We observe for any X , Y

$$X \wedge (X \equiv Y)$$
$$= \quad \{(27) \text{ with } Z := X \ \}$$
$$X \wedge X \equiv X \wedge Y \equiv X$$
$$= \quad \{\text{idempotence of } \wedge \ \}$$
$$X \equiv X \wedge Y \equiv X$$
$$= \quad \{\text{identity element of } \equiv \ \}$$
$$X \wedge Y \qquad .$$

(*End of Proof.*)

Remark on Notation When introducing the conjunction, we said that we would give it the same binding power as the disjunction, and that we did so for reasons of symmetry. By now we have seen a lot of symmetry. There is even more of it. We have, for instance, for any X , Y , Z —as the reader may care to verify—

$$[(X \wedge Y) \vee (Y \wedge Z) \vee (Z \wedge X) \equiv$$
$$(X \vee Y) \wedge (Y \vee Z) \wedge (Z \vee X)] \qquad .$$

With different binding powers for $\wedge$ and $\vee$, the parentheses in one of the lines are superfluous; omitting them would display the symmetry less nicely. We would like to stress that our decision to give $\wedge$ and $\vee$ the same binding power is based on much more than mere aesthetics; it is, in fact, based on a frightening observation.

In electronic engineering it is not unusual to associate the pair $\{0,1\}$ with the boolean scalar domain, in particular to associate the integer 1 with the boolean scalar *true* . Conjunction is then associated with multiplication and denoted by juxtaposition. This convention captures the idempotence of the conjunction: $xx = x$ does hold for $x = 0$ and $x = 1$; that 1 stands for the conjunction's identity element is captured by the arithmetically familiar $1x = x$. The next step is to denote the disjunction by $+$; its idempotence is rendered by $x + x = x$, which holds for $x = 0$; for $x = 1$, the analogy with arithmetic is violated. It is, however, perfect in the expression of the conjunction's distribution over disjunction:

$$x(y + z) = xy + xz \qquad ,$$

a relation with which all electronic engineers are thoroughly familiar. Our frightening observation was that many of them are more than hesitant about the dual

$$x + yz = (x + y)(x + z) \qquad .$$

(Try the experiment! Try to write down this last formula, and you will notice that it requires a conscious effort.)

This small example shows in its full horror how an unfortunate notation can damage one's manipulative abilities. We took it as a warning and decided that the destruction of symmetry in order to save a few characters is penny-wise and pound-foolish. (*End of Remark on Notation.*)

So much for the conjunction.

$$* \quad * \quad *$$

We continue our exploration by investigating (16), the other two-variable expression we could construct from $\equiv$ and $\vee$. This investigation requires some groundwork concerning punctuality. Because definition (3, 6) of punctuality uses the implication sign and we wish to avoid manipulating expressions containing the implication sign before the latter has been introduced in our calculus of boolean structures, we give here an alternative definition of punctuality. We shall see later that this definition of punctuality is equivalent to the earlier (3, 6) —see (46)— .

In this section, we use x , y , z to denote structures of some type. A function f from structures to structures being punctual means that for any x , y

(30) $[x = y \wedge f.x = f.y \equiv x = y]$.

Equality, besides being defined to be reflexive and symmetric, is postulated to be punctual in each of its operands, i.e., for any x , y , z —substitute in (30) $u = z$ for $f.u$ —

(31) $[x = y \wedge (x = z \equiv y = z) \equiv x = y]$.

Remark For the special case of the equivalence, the punctuality can be proved, thanks to the associativity of the equivalence. For the general equality, however, we had to postulate the punctuality. (*End of Remark.*)

Punctuality of the equality can be equivalently expressed by the property —sometimes known as "transitivity of $=$ "— that for any x , y , z

(32) $[x = y \wedge x = z \equiv x = y \wedge y = z]$.

Proof We observe for any x , y , z

$[x = y \wedge x = z \equiv x = y \wedge y = z]$ (i.e., (32))
$= \quad \{(27)$ with X , Y , $Z := x = y$, $x = z$, $y = z$ $\}$
$[x = y \wedge (x = z \equiv y = z) \equiv x = y]$ (i.e., (31)).
(*End of Proof.*)

Lemma The constant function is punctual, i.e., we have for any structures x , y , z

$$[x = y \land z = z \equiv x = y]$$.

Proof Follows immediately from the reflexivity of $=$ —i.e., $[z = z \equiv true]$— and the fact that *true* is the identity element of $\land$.

(*End of Proof.*)

Lemma The identity function is punctual, i.e., we have for any structures x , y

$$[x = y \land x = y \equiv x = y]$$.

Proof Follows immediately from the idempotence of $\land$.

(*End of Proof.*)

Lemma Let f , g , and the infix $\cdot$ be punctual in their arguments; then $f.x \cdot g.x$ is punctual in x , i.e., we have for any x , y

$$[x = y \land f.x \cdot g.x = f.y \cdot g.y \equiv x = y]$$.

Proof We observe for any x , y , and punctual f , g , and $\cdot$

$$x = y \land f.x \cdot g.x = f.y \cdot g.y$$
$=$ { g is punctual}
$$x = y \land g.x = g.y \land f.x \cdot g.x = f.y \cdot g.y$$
$=$ { $\cdot$ is punctual in its last argument}
$$x = y \land g.x = g.y \land f.x \cdot g.x = f.x \cdot g.y \land f.x \cdot g.x = f.y \cdot g.y$$
$=$ {transitivity of $=$ }
$$x = y \land g.x = g.y \land f.x \cdot g.x = f.x \cdot g.y \land f.x \cdot g.y = f.y \cdot g.y$$
$=$ { $\cdot$ is punctual in its last argument}
$$x = y \land g.x = g.y \land f.x \cdot g.y = f.y \cdot g.y$$
$=$ { g is punctual}
$$x = y \land f.x \cdot g.y = f.y \cdot g.y$$
$=$ { f is punctual}
$$x = y \land f.x = f.y \land f.x \cdot g.y = f.y \cdot g.y$$
$=$ { $\cdot$ is punctual in its first argument}
$$x = y \land f.x = f.y$$
$=$ { f is punctual}
$$x = y$$.

(*End of Proof.*)

The above proof has been conducted for a $\cdot$ with two arguments; from the structure of the proof we see that the conclusion holds for functions of any finite number of arguments. (The infix notation, chosen above in order to reduce the number of parentheses, is then no longer appropriate.)

With the above three lemmata, of which the last one takes care of induction over the grammar, we now conclude the

Punctuality Theorem Expressions built from variables and punctual operators are punctual functions of the variables.

Returning to boolean structures, we have

Lemma The disjunction is a punctual function of each of its arguments, i.e., for any boolean structures X , Y , Z

$$[(X \equiv Y) \wedge (X \vee Z \equiv Y \vee Z) \equiv (X \equiv Y)] \qquad .$$

Proof We observe for any X , Y , Z

$(X \equiv Y) \wedge (X \vee Z \equiv Y \vee Z)$
= $\{$ $\vee$ distributes over $\equiv$ $\}$
$(X \equiv Y) \wedge ((X \equiv Y) \vee Z)$
= $\{$Law of Absorption$\}$
$(X \equiv Y)$.

$\qquad\qquad\qquad\qquad\qquad\qquad\qquad$ (*End of Proof.*)

From the Punctuality Theorem and the punctuality of $\equiv$ and $\vee$ we immediately conclude that —see the Golden Rule— $\wedge$ is punctual as well. Furthermore we conclude that $X \vee Y \equiv Y$, the expression we had set out to investigate, is a punctual function of X and Y . It has, indeed, more nice properties, as follows from the little theory given below; we give that little theory in isolation because it is perfectly general.

In presenting the little theory, we have taken the nonpuritan decision to use the implication despite the fact that it has not yet been introduced in our calculus of boolean structures. We took that decision for two reasons. Firstly, in normal mathematical parlance the implication occurs in the statement of many of the properties the little theory is about. Secondly, without the implication sign our proofs would have been much longer. Because the little theory is perfectly general, we prefer to present the proofs in the conciseness we are used to and to use implication as we see fit. So, for a punctual function f of the proper type we use formulae like

(33) $[x = y \wedge f.x \Rightarrow f.y]$

(34) $[x = y \wedge f.x \Rightarrow f.x \wedge f.y]$

(35) $[x = y \Rightarrow f.x = f.y]$.

(Here $\Rightarrow$ has the lowest binding power.)

Consoling Remark The puritan reader who is too much alarmed by our lack of orthodoxy has two options. Either he can reformulate our little theory and thus convince himself that our premature use of the implication is no more than an abbreviation. Or he can ignore our general little theory altogether and demonstrate directly all the properties of the expression $X \vee Y \equiv Y$ that we derive from it. (*End of Consoling Remark.*)

Little Theory Let x , y , z be structures of some type. Consider a punctual infix operator $\cdot$ and a binary relation $\rightarrow$ that is defined in terms of $\cdot$ by

(36) $[x \rightarrow y \equiv x \cdot y = y]$.

(In this Little Theory, the order of decreasing binding power of the operators is

$$\infty \;,\; \cdot \;,\; \rightarrow \;\text{ and }\; = \;,\; \wedge \;\text{ and }\; \vee \;,\; \Rightarrow \;,\; \equiv \;.)$$

Theorem ($\cdot$ is idempotent) $\equiv$ ($\rightarrow$ is reflexive)   .

Proof We observe for any x

$\quad [x \cdot x = x]$
$=\quad \{(36) \text{ with } y := x \}$
$\quad [x \rightarrow x]$,

from which observation the theorem follows. (*End of Proof.*)

Theorem ($\cdot$ is associative) $\Rightarrow$ ($\rightarrow$ is transitive)   .

Proof Because the consequent states for any x , y , z

$$[x \rightarrow y \wedge y \rightarrow z \Rightarrow x \rightarrow z]\qquad,$$

we observe for any x , y , z

$\quad x \rightarrow y \wedge y \rightarrow z$
$=\quad \{(36); (36) \text{ with } x , y := y , z \}$
$\quad x \cdot y = y \wedge y \cdot z = z$
$\Rightarrow\quad \{(34) \text{ with } x , y := y , x \cdot y; \text{ Punctuality Theorem for } f \text{ given by}$
$\qquad [f.u \equiv u \cdot z = z] \}$
$\quad y \cdot z = z \wedge (x \cdot y) \cdot z = z$
$=\quad \{ \cdot \text{ is associative}\}$
$\quad y \cdot z = z \wedge x \cdot (y \cdot z) = z$
$\Rightarrow\quad \{(33) \text{ with } x , y := y \cdot z , z; \text{ Punctuality Theorem for } f \text{ given by}$
$\qquad [f.u \equiv x \cdot u = z] \}$
$\quad x \cdot z = z$
$=\quad \{(36) \text{ with } y := z \}$
$\quad x \rightarrow z$.

(*End of Proof.*)

Theorem ($\cdot$ is symmetric) $\Rightarrow$ ($\rightarrow$ is antisymmetric)   .

Proof Because the consequent states for any x , y

$$[x \rightarrow y \wedge y \rightarrow x \Rightarrow x = y]$$,

we observe for any x , y

$\quad x \rightarrow y \wedge y \rightarrow x$
$=\quad \{(36); (36)$ with x , $y := y$, x $\}$
$\quad x \cdot y = y \wedge y \cdot x = x$
$=\quad \{$ $\cdot$ is symmetric$\}$
$\quad x \cdot y = y \wedge x \cdot y = x$
$\Rightarrow\quad \{$ $=$ is transitive$\}$
$\quad x = y$.

                                            (*End of Proof.*)

Theorem With ∞ a punctual prefix operator,

(∞ distributes over $\cdot$) $\Rightarrow$ (∞ is monotonic with respect to $\rightarrow$)   .

Proof Because the consequent states that for any x , y

$$[x \rightarrow y \Rightarrow \infty x \rightarrow \infty y]$$,

we observe for any x , y

$\quad \infty x \rightarrow \infty y$
$=\quad \{(36)$ with $x, y := \infty x, \infty y$ $\}$
$\quad \infty x \cdot \infty y = \infty y$
$=\quad \{$ ∞ distributes over $\cdot$ $\}$
$\quad \infty(x \cdot y) = \infty y$
$\Leftarrow\quad \{(35)$ with $f. := \infty$, which is punctual$\}$
$\quad x \cdot y = y$
$=\quad \{(36)\}$
$\quad x \rightarrow y$.

                                            (*End of Proof.*)

Since an operator that is idempotent, associative, and symmetric —see (14) and its proof— enjoys the property of "auto-distribution", we have the

Corollary ($\cdot$ is idempotent, associative, and symmetric) $\Rightarrow$
                ($\cdot$ is monotonic with respect to $\rightarrow$)   .

Theorem (1 is a left identity element of $\cdot$) $\equiv$
                (1 is a left extreme of $\rightarrow$)   .

Proof We observe for any y

$$1 \cdot y = y$$
$$= \quad \{(36) \text{ with } x := 1 \}$$
$$1 \rightarrow y \quad ,$$

from which observation the theorem follows.

(*End of Proof.*)

Theorem (0 is a right zero element of $\cdot$) $\equiv$
(0 is a right extreme of $\rightarrow$)   .

Proof We observe for any x

$$x \cdot 0 = 0$$
$$= \quad \{(36) \text{ with } y := 0 \}$$
$$x \rightarrow 0 \quad ,$$

from which observation the theorem follows.

(*End of Proof.*)

(*End of Little Theory.*)

Since $\vee$ is idempotent, associative, and symmetric, the above little theory tells us that $X \vee Y \equiv Y$ is, indeed, a nice expression. The corresponding relation is called the implication, it is denoted by $\Rightarrow$, and read as "at weakest" or as "implies". It is defined by the postulate

(37) $[X \Rightarrow Y \equiv X \vee Y \equiv Y]$;

we give $\Rightarrow$ a higher binding power than $\equiv$ but a lower binding power than $\wedge$ and $\vee$. We identify it with the implication as we have used it all the time. Because it is not symmetric, its operands need different names; in $X \Rightarrow Y$, X is called the antecedent and Y the consequent.

From (37) and the Punctuality Theorem it follows that the implication is a punctual function of its arguments. Furthermore, the little theory tells us

• because $\vee$ is idempotent, $\Rightarrow$ is reflexive:

(38) $[X \Rightarrow X]$;

• because $\vee$ is associative, $\Rightarrow$ is transitive:

(39) $[(X \Rightarrow Y) \wedge (Y \Rightarrow Z) \Rightarrow (X \Rightarrow Z)]$;

• because $\vee$ is symmetric, $\Rightarrow$ is antisymmetric:

(40) $[(X \Rightarrow Y) \wedge (Y \Rightarrow X) \Rightarrow (X \equiv Y)]$;

- because $\lor$ is idempotent, associative, and symmetric, $\lor$ is monotonic with respect to $\Rightarrow$:

$$(41) \qquad [(X \Rightarrow Y) \Rightarrow (X \lor Z \Rightarrow Y \lor Z)] \qquad ;$$

- because *true* is the (right) zero element of $\lor$, it is the right extreme of $\Rightarrow$:

$$(42) \qquad [X \Rightarrow true] \qquad .$$

Relation (40) underlies the ping-pong argument that establishes equivalence by mutual implication. We have, in fact, for any X , Y

$$(43) \qquad [(X \Rightarrow Y) \land (Y \Rightarrow X) \;\equiv\; (X \equiv Y)] \qquad ,$$

which we prove directly.

Proof We observe for any X , Y :

$$(X \Rightarrow Y) \land (Y \Rightarrow X)$$
$$= \quad \{(37); (37) \text{ with } X, Y := Y, X \}$$
$$(X \lor Y \equiv Y) \land (Y \lor X \equiv X)$$
$$= \quad \{ \lor \text{ is symmetric}\}$$
$$(X \lor Y \equiv Y) \land (X \lor Y \equiv X)$$
$$= \quad \{ \equiv \text{ is transitive}\}$$
$$(X \equiv Y) \land (X \lor Y \equiv X)$$
$$= \quad \{ \lor \text{ and } \equiv \text{ are punctual}\}$$
$$(X \equiv Y) \land (X \lor X \equiv X)$$
$$= \quad \{ \lor \text{ is idempotent}\}$$
$$(X \equiv Y) \land (X \equiv X)$$
$$= \quad \{\text{identity element of } \equiv \}$$
$$(X \equiv Y) \land true$$
$$= \quad \{\text{identity element of } \land \}$$
$$(X \equiv Y) \qquad .$$

(End of Proof.)

Because also the conjunction is idempotent, associative, and symmetric, it stands to reason to introduce the consequence, denoted by $\Leftarrow$ —with the same binding power as $\Rightarrow$ — and read as "at strongest" or as "follows from". It is defined by the postulate

$$(44) \qquad [X \Leftarrow Y \;\equiv\; X \land Y \;\equiv\; Y] \qquad .$$

For reasons of symmetry, the consequence is a nice thing to have, but it gives us nothing new, as shown by the following

$$(45) \qquad [X \Rightarrow Y \equiv Y \Leftarrow X] \qquad .$$

Proof We observe for any X , Y

$\quad X \Rightarrow Y$
$= \quad \{(37)\}$
$\quad X \vee Y \equiv Y$
$= \quad \{\text{Golden Rule}\}$
$\quad Y \wedge X \equiv X$
$= \quad \{(44) \text{ with } X,Y := Y,X \}$
$\quad Y \Leftarrow X$.

$\hfill$ (*End of Proof.*)

It is useful to know the corollary from (44) and (45)

(46) $\qquad [X \Rightarrow Y \equiv X \wedge Y \equiv X]$

and the conclusion that also the conjunction is monotonic with respect to the implication. The last formula comes in handy to prove

(47) $\qquad [(X \Rightarrow Y) \wedge (X' \Rightarrow Y') \Rightarrow (X \wedge X' \Rightarrow Y \wedge Y')]$.

Proof We observe for any X , X' , Y , Y'

$\quad (X \Rightarrow Y) \wedge (X' \Rightarrow Y')$
$= \quad \{(46), \text{ twice}\}$
$\quad (X \wedge Y \equiv X) \wedge (X' \wedge Y' \equiv X')$
$\Rightarrow \quad \{ \wedge \text{ and } \equiv \text{ are punctual}\}$
$\quad (X \wedge Y) \wedge (X' \wedge Y') \equiv X \wedge X'$
$= \quad \{(46)\}$
$\quad X \wedge X' \Rightarrow Y \wedge Y'$.

$\hfill$ (*End of Proof.*)

The original definition (37) of the implication can be used similarly to demonstrate

(48) $\qquad [(X \Rightarrow Y) \wedge (X' \Rightarrow Y') \Rightarrow (X \vee X' \Rightarrow Y \vee Y')]$.

It is useful to know the following reformulation of the Laws of Absorption; (23) and (24) yield in that order

(49) $\qquad [X \Rightarrow X \vee Y]$

(50) $\qquad [X \wedge Y \Rightarrow X]$.

For an equivalence as consequent we have

Theorem For any X , Y , Z

(51) $[X \Rightarrow (Y \equiv Z) \; \equiv \; X \wedge Y \; \equiv \; X \wedge Z]$.

Proof We observe for any X , Y , Z

$X \Rightarrow (Y \equiv Z)$
= $\{(46) \text{ with } Y := (Y \equiv Z) \}$
   $X \wedge (Y \equiv Z) \equiv X$
= $\{(27)\}$
   $X \wedge Y \equiv X \wedge Z$.

(*End of Proof.*)

Consequents of the form of an equivalence being not unusual, the above theorem is one to remember. We shall refer to it later in this chapter.

*   *   *

It is time to introduce a really new operator, the unary prefix operator called negation, written as "$\neg$" , and read as "non"; we give it a higher binding power than $\vee$ and $\wedge$. Negation is a function from boolean structures to boolean structures; applied to a boolean scalar, the negation yields a boolean scalar, i.e., $\neg true$ is a boolean scalar, or, formally, —see (1)— we have

$$[\neg true] \equiv \neg true   .$$

It is customary to denote the boolean scalar $\neg true$ by *false* , i.e., we have

(52) $false \equiv \neg true$

(53) $[false] \equiv false$.

The properties of *false* will be derived from those of the negation.

We postulate separately the properties of the negation with respect to the equivalence and the disjunction, i.e., the operators we have treated as the primary ones. From those postulates we shall derive its properties with respect to the secondary ones, conjunction and implication.

Negation and equivalence are connected by the postulate that for any X , Y

(54) $[\neg(X \equiv Y) \equiv \neg X \equiv Y]$.

Theorem For any X

(55) $[\neg X \equiv X \equiv false]$.

Proof We observe for any X

$\quad \neg X \equiv X$
$=\quad \{(54)$ with $Y \ := X\ \}$
$\quad \neg(X \equiv X)$
$=\quad \{(7)\}$
$\quad \neg true$
$=\quad \{(52)\}$
$\quad false$.

                                        (*End of Proof.*)

Theorem For any X , Y

(56) $[\neg X \equiv Y \equiv X \equiv \neg Y]$.

Proof We observe for any X , Y

$\quad \neg X \equiv Y$
$=\quad \{(54)\}$
$\quad \neg(X \equiv Y)$
$=\quad \{$symmetry of $\equiv$; (54) with X , $Y \ := Y$, $X\ \}$
$\quad X \equiv \neg Y$.

                                        (*End of Proof.*)

Theorem   Negation is its own inverse, i.e., for any X

(57) $[\neg\neg X \equiv X]$.

Proof We observe for any X

$\quad \neg\neg X \equiv X$
$=\quad \{(56)$ with $Y := \neg X\ \}$
$\quad \neg X \equiv \neg X$
$=\quad \{$identity element of $\equiv\ \}$
$\quad true$.

                                        (*End of Proof.*)

Theorem   The negation is a punctual function of its argument, i.e., for any X , Y

(58) $[(X \equiv Y) \wedge (\neg X \equiv \neg Y) \equiv (X \equiv Y)]$.

Proof We observe for any X , Y

$\quad (X \equiv Y) \wedge (\neg X \equiv \neg Y)$

$= \quad \{(56)\}$

$\quad (X \equiv Y) \wedge (X \equiv Y)$

$= \quad \{ \ \wedge \ \text{ is idempotent}\}$

$\quad X \equiv Y \qquad .$

$\hfill$ (*End of Proof.*)

Negation and disjunction are connected by the famous Law of the Excluded Middle. We postulate for any X

(59) $\qquad [X \vee \neg X] \qquad .$

In order to investigate what we can derive from (54) and (59) together, we observe that the latter contains one negation sign, whereas the former gives an alternative expression for the negated equivalence. Thus inspired, we observe for any X , Y

$\quad true$

$= \quad \{(59) \text{ with } X := (X \equiv Y) \ \}$

$\quad [(X \equiv Y) \vee \neg(X \equiv Y)]$

$= \quad \{(54)\}$

$\quad [(X \equiv Y) \vee (\neg X \equiv Y)]$

$= \quad \{ \ \vee \ \text{ distributes over } \equiv \ \}$

$\quad [X \vee \neg X \equiv Y \vee \neg X \equiv X \vee Y \equiv Y \vee Y]$

$= \quad \{(59) \text{ and identity element of } \equiv \ \}$

$\quad [Y \vee \neg X \equiv X \vee Y \equiv Y \vee Y]$

$= \quad \{\text{properties of } \vee \ \}$

$\quad [\neg X \vee Y \equiv X \vee Y \equiv Y] \qquad .$

Thus we have proved the

Theorem For any X , Y

(60) $\qquad [\neg X \vee Y \equiv X \vee Y \equiv Y] \qquad .$

Remark For the variation we chose to explore a calculational opportunity, just to see what theorem we would come up with. Platonists would say that we have "discovered" a new theorem. (*End of Remark.*)

That is a nice theorem, for the substitution $X := true$ gives us a handle on the rôle of *false* as a disjunct. Indeed we have the

Theorem The boolean scalar *false* is the identity element of the disjunction, i.e., for any X

(61) $[X \vee false \equiv X]$.

Proof We observe for any X

$\quad X \vee false$
$= \quad \{(52)\}$
$\quad X \vee \neg true$
$= \quad \{(60)$ with $X, Y := true, X$ $\}$
$\quad true \vee X \equiv X$
$= \quad \{$ $true$ is zero element of $\vee$ $\}$
$\quad true \equiv X$
$= \quad \{$ $true$ is identity element of $\equiv$ $\}$
$\quad X$.

(*End of Proof.*)

And in the same vein we have the

Theorem The boolean scalar *false* is the zero element of the conjunction, i.e., for any X

(62) $[X \wedge false \equiv false]$.

Proof We observe for any X

$\quad X \wedge false \equiv false$
$= \quad \{$Golden Rule$\}$
$\quad X \equiv X \vee false$
$= \quad \{(61)\}$
$\quad true$.

(*End of Proof.*)

Formula (60) has not been exhausted yet. It gives us a way of eliminating the negation from a negated disjunct. Let us see what happens if we start with two negated disjuncts and apply (60) twice! We observe for any X , Y

$$\neg X \vee \neg Y$$
$$= \quad \{(60) \text{ with } Y := \neg Y \}$$
$$X \vee \neg Y \equiv \neg Y$$
$$= \quad \{(60) \text{ with } X, Y := Y, X \}$$
$$Y \vee X \equiv X \equiv \neg Y$$
$$= \quad \{(54) \text{ with } X, Y := Y , Y \vee X \equiv X \}$$
$$\neg (Y \vee X \equiv X \equiv Y)$$
$$= \quad \{\text{Golden Rule}\}$$
$$\neg (X \wedge Y) \qquad .$$

Thus we have derived one of the well-known Laws of Augustus de Morgan, viz., that for any X , Y

(63) $[\neg X \vee \neg Y \equiv \neg (X \wedge Y)]$.

Eliminating from (60) the two disjunctions by means of the Golden Rule yields after simplification

(64) $[\neg X \wedge Y \equiv X \wedge Y \equiv \neg Y]$,

a rewrite rule to remove the negation from a negated conjunct. Applying it twice provides one of the many ways of deriving the other Law of de Morgan, viz., that for any X , Y

(65) $[\neg X \wedge \neg Y \equiv \neg (X \vee Y)]$.

We leave to the reader the verification that the Laws of de Morgan also hold for more than two dis- and conjuncts, e.g.,

$$[\neg X \vee \neg Y \vee \neg Z \equiv \neg (X \wedge Y \wedge Z)] \qquad .$$

Remark Ironically, de Morgan lacked the proper notational tools to express the laws he is best known for: he did not have the negation operator. Knowing that the negation is its own inverse and that the 26 letters of the upper-case alphabet are in one-to-one correspondence to the 26 letters from the lower-case alphabet, he introduced the convention that the two cases of the same letter should stand for a proposition and its negation. Consequently he could only negate named propositions and not expressions (which are what his Law is about). He had to write something like

$$(x \wedge y) = z \equiv (X \vee Y) = Z \qquad .$$

At first sight, de Morgan's convention might seem a neat way of capturing that negation is its own inverse; we would like to stress that, for its lack of combinatorial freedom, it is from the point of view of manipulation a severe pain in the neck. (*End of Remark*.)

A very different theorem connecting negation, disjunction, and conjunction is the following equivalence. It may strike the reader as a rather special theorem; it has been included because it equates expressions of a form we shall encounter frequently when dealing with the semantics of the repetition.

Theorem For any X , Y , Z

(66) $[(\neg X \vee Y) \wedge (X \vee Z) \equiv (X \wedge Y) \vee (\neg X \wedge Z)]$.

Proof We observe for any X , Y

   $(\neg X \vee Y) \wedge (X \vee Z)$
= {because it enables us to form the conjunctions at the right-hand side of
   (66), we distribute $\wedge$ over $\vee$ }
   $(\neg X \wedge X) \vee (Y \wedge X) \vee (\neg X \wedge Z) \vee (Y \wedge Z)$
= {we must eliminate the outer disjuncts; the left one is easy: with de
   Morgan and the Excluded Middle $[\neg X \wedge X \equiv false]$, and *false* is
   the identity element of $\vee$ }
   $(X \wedge Y) \vee (\neg X \wedge Z) \vee (Y \wedge Z)$
= {heading for the Law of Absorption, we introduce X and $\neg X$ into the
   last disjunct; with the Excluded Middle and the identity element of $\wedge$ }
   $(X \wedge Y) \vee (\neg X \wedge Z) \vee ((X \vee \neg X) \wedge Y \wedge Z)$
= {distribute $\wedge$ over $\vee$ }
   $(X \wedge Y) \vee (X \wedge Y \wedge Z) \vee (\neg X \wedge Z) \vee (\neg X \wedge Z \wedge Y)$
= {Law of Absorption, twice}
   $(X \wedge Y) \vee (\neg X \wedge Z)$.

(*End of Proof.*)

So much for the conjunction. We now turn our attention to the connections between negation and implication; (60) is not exhausted yet.

Theorem For any X , Y

(67) $[X \Rightarrow Y \equiv \neg X \vee Y]$.

Proof We observe for any X , Y

   $X \Rightarrow Y$
= {(37)}
   $X \vee Y \equiv Y$
= {(60)}
   $\neg X \vee Y$.

(*End of Proof.*)

It has as immediate offspring what is known as the

Shunting Theorem For any X , Y , Z

(68) $[X \wedge Y \Rightarrow Z \equiv X \Rightarrow \neg Y \vee Z]$.

Proof We observe for any X , Y , Z

$X \wedge Y \Rightarrow Z$
=   {(67) with $X,Y := (X \wedge Y),\ Z$ }
$\neg(X \wedge Y) \vee Z$
=   {de Morgan}
$\neg X \vee \neg Y \vee Z$
=   {(67) with $Y := \neg Y \vee Z$ }
$X \Rightarrow \neg Y \vee Z$.

                                                                (*End of Proof.*)

A pure connection between negation and implication is given by the

Theorem of the Contra-positive For any X , Y

(69) $[X \Rightarrow Y \equiv \neg Y \Rightarrow \neg X]$.

Proof We observe for any X , Y

$\neg Y \Rightarrow \neg X$
=   {(67) with $X,Y := \neg Y,\ \neg X$ }
$\neg\neg Y \vee \neg X$
=   { $\neg$ is its own inverse}
$\neg X \vee Y$
=   {(67)}
$X \Rightarrow Y$.

                                                                (*End of Proof.*)

The proof of

(70) $[\textit{false} \Rightarrow X]$ for all X

is left to the reader.

For the sake of completeness, we mention yet another infix operator, called the discrepancy, written as $\neq$, and read as "differs from". It is symmetric and associative; it is mutually associative with the equivalence and has been given the same low binding power as the equivalence. It is defined by

(71) $[X \neq Y \equiv \neg(X \equiv Y)]$.

We leave to the reader the verification of

(72) $[X \not\equiv false \equiv X]$

(73) $[X \wedge (Y \not\equiv Z) \equiv X \wedge Y \not\equiv X \wedge Z]$.

The fact that conjunction distributes over discrepancy is about the only reason for mentioning the discrepancy at all.

Remark   Nothing in our postulates prevents us from choosing for the negation the identity operator. From the Excluded Middle we then immediately derive, for any X , $[X]$ —or $[X \equiv true]$, if we wish to be more explicit— . There is nothing logically wrong with this model, in which *true* is essentially the only boolean structure; the model's only disadvantage is that it is totally void of interest. Hence our interest is restricted to those models in which $[X]$ holds only for X the boolean scalar *true* and for none of the other boolean structures, of which at least one exists. The rejection of the noninteresting model is more than we care to formalize. (*End of Remark.*)

*   *   *

We need one further postulate about the "everywhere" operator, which we introduced as a function from boolean structures to the traditional boolean domain $\{true, false\}$. We have seen that

(i)   it is idempotent,
(ii) it has *true* and *false* as fixpoints, i.e.,

$$[true] = true \qquad \text{and} \qquad [false] = false \qquad .$$

As we shall see later in this chapter, (i) and (ii) are properties enjoyed by quantification over a non-empty range, be it existential or universal quantification. In order to give the "everywhere" operator the properties of universal quantification, we postulate that it distributes over conjunction, i.e., for any X , Y

(74) $[X \wedge Y] \equiv [X] \wedge [Y]$.

As a consequence, it is monotonic with respect to the implication, i.e., for any X , Y

(75) $[X \Rightarrow Y] \Rightarrow ([X] \Rightarrow [Y])$.

*   *   *

In the above, we have gone in great detail through a few dozen formulae. Some might even argue that we spent more pages on them than the topic deserves. This, however, is not confirmed by the general experience (of us and of others). In the teaching of this material, two handicaps are quite common.

The one handicap consists in an audience of people that have used boolean expressions in programming or in circuit design and therefore believe that they know all these things already. They have, indeed, seen some of the operators, but their knowledge is usually rather incomplete, and sometimes even twisted. (Famous is the story of the electronic engineer that knew the discrepancy, which he called "the exclusive or", but had never thought of the equivalence; when asked to do the latter, he came up with "the exclusive nor".) The operators they know, they usually know by means of "truth tables", and extensive case analysis —i.e., substituting all possible values for all variables— is often their only way of coping with boolean expressions. With increasing number of variables, this way becomes very clumsy, and, in the presence of nonpunctual functions or quantified expressions, it is fundamentally inadequate. We are much better off with the rules of manipulation at our fingertips, but these rules are almost totally unknown. Those rules are what we wish to convey in this chapter.

The other handicap consists in an audience that has been introduced to logic by philosophers. Such an audience can become very uneasy about our use of the associativity of the equivalence, because that use does not reflect human reasoning, which, according to some philosophers, logic has to mimic. Without denying the associativity of the equivalence, people can violently protest against its exploitation because it is "unnatural" or "counter-intuitive".

We grant the latter in the sense that (at least Western) languages are rather ill-equipped at expressing equivalence. We have, of course, the infamous "if and only if" —where "if" takes care of "follows from" and "only if" of "implies"— but that is no more than an unfortunate patch: by all linguistic standards, the sentence

"Tom can see with both eyes if and only if Tom can see with only one eye if and only if Tom is blind."

is —probably for its blatant syntactic ambiguity— total gibberish. But for us this is no reason to disqualify the equivalence. On the contrary, if our formalism allows simple calculations that are beyond the unaided mind because their verbal rendering would be too baffling, so much the better. In this respect we are totally pragmatic, and it was in that vein that we stressed the algebraic nature of the calculus.

So much for the two handicaps. There is a potential third one, viz., the possible inclination to reinterpret all the time our formulae in set-theoretic terms.

A possible model for our boolean structures is provided by the boolean functions defined on some space, with all punctual operators being applied point-wise. The next step is to establish a one-to-one correspondence between the subsets of points from the space and the boolean structures: with each subset of points we associate its membership function as its corresponding boolean structure.

We can now try to translate our logical operators into set-theoretic terms. For disjunction and conjunction, this is easy: they correspond to union $\cup$ and intersection $\cap$, respectively. With the equivalence we have the problem that (for two operands) it would yield the set of points belonging to both operands or to neither, and, traditionally, set theory shuns operators that yield sets containing elements that don't belong to at least one of the operands. Fortunately, there is what is called "the symmetric set difference" $\div$: it contains all elements belonging to one operand but not to the other. It corresponds to our discrepancy $\neq$, and in set theory the associativity of the symmetric set difference is known. The equality sign $=$ is used to denote equality of subsets; it corresponds to our "equivales everywhere" $[\equiv]$.

In order to render the Golden Rule in standard set-theoretical notation, we have to replace two equivalences by discrepancies, whereas the last one is replaced by $=$ to carry the burden of the square brackets. Even in serious books on set theory we may thus find the following "different" theorems

$$A = B \div (A \cap B) \div (A \cup B)$$
$$A \div B = (A \cap B) \div (A \cup B)$$
$$(A \cap B) = A \div B \div (A \cup B)$$
$$(A \cup B) = A \div B \div (A \cap B)$$
$$A \div (A \cup B) = B \div (A \cap B) \qquad ,$$

in which we recognize five clumsy renderings of the same Golden Rule. The above is a striking example of how inadequate notation can generate spurious diversification. The main culprit, of course, is not the symmetric set difference —this could be remedied by the introduction of a "symmetric set equality"— but the absence of the square brackets. We hope that the above reinforces our recommendation not to translate our formulae into set-theoretical notation and concepts so as to make them "easier to understand". Besides the clumsiness of that translation, there is a more fundamental reason for not indulging in it: the model of the subsets is overspecific because the analogue of their elements need not exist.

$$* \qquad * \qquad *$$

As the reader will have noticed, we did not overstress the implication. We encountered a whole bunch of equivalent expressions, such as

$$X \vee Y \equiv Y$$
$$X \wedge Y \equiv X$$
$$\neg X \vee Y$$
$$\neg X \wedge \neg Y \equiv \neg Y$$
$$\neg X \vee \neg Y \equiv \neg X$$
$$\neg (X \wedge \neg Y) \qquad .$$

Faced with such an *embarras du choix*, one should consider the introduction of a neutral shorthand for them. So we did: we introduced $X \Rightarrow Y$ for all of them. As soon as one has done so, one is invited to study the properties of the newly introduced relation. We found with our little theory that it is reflexive, transitive, and antisymmetric and that disjunction and conjunction are monotonic with respect to it. So far, so good.

A punctual function f being antimonotonic with respect to the implication means that for any X , Y

$$[(X \Rightarrow Y) \Rightarrow (f.Y \Rightarrow f.X)]$$.

The Law of the Contra-positive tells us that negation is antimonotonic; (67) tells us that implication is both: monotonic in its consequent and antimonotonic in its antecedent. And equivalence is neither.

One could try to use maintenance of implications as one's primary proof paradigm, an implication being maintained by strengthening its antecedent or by weakening its consequent. Because of one's reliance on monotonicity properties, an equivalence then becomes an awkward expression to manipulate. It is customary to replace it by mutual implication:

$$[X \equiv Y \equiv (X \Rightarrow Y) \wedge (X \Leftarrow Y)]$$.

If this is one's only way of eliminating the equivalence, the price is very heavy: eliminating the equivalences from $X \equiv Y \equiv Z$ by mutual implication irrevocably destroys the symmetry.

Using monotonicity with respect to implication is one thing; manipulating expressions that make heavy use of the implication is quite another matter. Firstly, the implication is neither symmetric nor associative, and its laws of distribution are awkward. Secondly, in combination with the negation sign, hell breaks loose, as is shown by the following eight expressions:

$$(\neg X \Rightarrow Y) \Rightarrow \neg(X \Rightarrow \neg Y)$$
$$(\neg Y \Rightarrow X) \Rightarrow \neg(X \Rightarrow \neg Y)$$
$$(\neg X \Rightarrow Y) \Rightarrow \neg(Y \Rightarrow \neg X)$$
$$(\neg Y \Rightarrow X) \Rightarrow \neg(Y \Rightarrow \neg X)$$
$$(X \Rightarrow \neg Y) \Rightarrow \neg(\neg X \Rightarrow Y)$$
$$(Y \Rightarrow \neg X) \Rightarrow \neg(\neg X \Rightarrow Y)$$
$$(X \Rightarrow \neg Y) \Rightarrow \neg(\neg Y \Rightarrow X)$$
$$(Y \Rightarrow \neg X) \Rightarrow \neg(\neg Y \Rightarrow X)$$,

which —believe it or not— are all equivalent to $X \equiv Y$. They are all derivatives of

(76) $$[X \equiv Y \equiv X \vee Y \Rightarrow X \wedge Y]$$.

Proof of (76) We observe for any X, Y

$\quad X \vee Y \Rightarrow X \wedge Y$
$=\quad$ {since —Law of Absorption— $[X \vee Y \Leftarrow X \wedge Y]$ }
$\quad (X \vee Y \Rightarrow X \wedge Y) \wedge (X \vee Y \Leftarrow X \wedge Y)$
$=\quad$ {(43), i.e., mutual implication}
$\quad X \vee Y \equiv X \wedge Y$
$=\quad$ {Golden Rule}
$\quad X \equiv Y \qquad .$

$\hfill$ (*End of Proof.*)

And all that glut was derived without using the Shunting Theorem! The moral of the story is that expressions with the implication sign offer too much needless variety, and that the corresponding arsenal of manipulation rules is larger (and uglier) than we care to remember. Consequently, the proper elimination of the implication sign —as in the above proof— is a usual first step in our calculations.

Remark In his connection we would like to draw attention to two observations.

Firstly, people thoroughly familiar with equivalence as mutual implication are often surprised by (76); this is strange, for (76) is —see the above proof— in a sense the dual of the mutual implication.

Secondly, people thoroughly familiar with the transitivity of $\leqslant$:

$$x \leqslant y \wedge y \leqslant z \Rightarrow x \leqslant z$$

are often surprised to see for its negation $>$:

$$x > z \Rightarrow x > y \vee y > z \qquad .$$

(Should we call this property of $>$ "antitransitivity"?) You are cordially invited to take the experiment yourself and to show to your random colleague the above formula with the question of whether it is a theorem.

Some of our habits are needlessly asymmetric. (*End of Remark.*)

$$*\quad * \quad *$$

It is time to introduce quantification. We shall introduce two forms, universal quantification and existential quantification. They are each other's dual. We shall introduce the universal quantification *in extenso* and shall then state how the duality gives the corresponding properties of the existential quantification.

Universal quantification is a generalization of conjunction. Its format is

$\quad (\forall \text{dummies: range: term}) \qquad .$

Here, "dummies" stands for an unordered list of local variables, whose scope is delineated by the outer parenthesis pair. In what follows, x and y will be used to denote dummies; the dummies may be of any understood type.

The two components "range" and "term" are boolean structures, and so is the whole quantified expression, which is a boolean scalar if both range and term are boolean scalars. Range and term may depend on the dummies; their potential dependence on the dummies will be indicated explicitly by using a functional notation, e.g., if a range has the form $r.x \land s.x.y$, it is a conjuntion of $r.x$, which may depend on x but does not depend on y , and $s.x.y$, which may depend on both. Similarly, X will stand for a boolean structure that depends on none of the dummies. In the following we shall use r , s , f , and g to denote functions from (the types of) the dummies to boolean structures.

For the sake of brevity, the range $true$ is omitted. The following postulate tells us how ranges different from $true$ can be eliminated:

(77) $[(\forall x:\ r.x:\ f.x) \equiv (\forall x::\ \neg r.x \lor f.x)]$.

(Appeals to this postulate will be given by the catchword "trading", being short for "trading between range and term".) The convenience of ranges differing from $true$ will transpire later.

Disjunction distributes in the same way over universal quantification as it does over conjunction, i.e., we postulate for any X , f

(78) $[X \lor (\forall x::\ f.x) \equiv (\forall x::\ X \lor f.x)]$.

But now we observe for any X , r , f

$\quad X \lor (\forall x:\ r.x:\ f.x)$
$=\quad \{\text{trading}\}$
$\quad X \lor (\forall x::\ \neg r.x \lor f.x)$
$=\quad \{(78)\text{ with } f.x := \neg r.x \lor f.x \ \}$
$\quad (\forall x::\ X \lor \neg r.x \lor f.x)$
$=\quad \{\text{trading}\}$
$\quad (\forall x:\ r.x:\ X \lor f.x)$.

Hence we have for any X , r , f

(79) $[X \lor (\forall x:\ r.x:\ f.x) \equiv (\forall x:\ r.x:\ X \lor f.x)]$.

In analogy to conjunction's symmetry and associativity, we postulate that universal quantification distributes over conjunction, i.e., we postulate for any f and g

(80) $[(\forall x::\ f.x) \land (\forall x::\ g.x) \equiv (\forall x::\ f.x \land g.x)]$,

which invites us to observe for any r , f , g

$$(\forall x:\ r.x:\ f.x) \wedge (\forall x:\ r.x:\ g.x)$$
$=$ {trading, twice}
$$(\forall x::\ \neg r.x \vee f.x) \wedge (\forall x::\ \neg r.x \vee g.x)$$
$=$ {(80) with $f.x$, $g.x := \neg r.x \vee f.x$, $\neg r.x \vee g.x$ }
$$(\forall x::\ (\neg r.x \vee f.x) \wedge (\neg r.x \vee g.x))$$
$=$ { $\vee$ distributes over $\wedge$ }
$$(\forall x::\ \neg r.x \vee (f.x \wedge g.x))$$
$=$ {trading}
$$(\forall x:\ r.x:\ f.x \wedge g.x) \qquad .$$

Hence $\forall$ distributes over $\wedge$, i.e., for any r , f , g

(81) $[(\forall x:\ r.x:\ f.x) \wedge (\forall x:\ r.x:\ g.x) \equiv (\forall x:\ r.x:\ f.x \wedge g.x)] \qquad .$

Formulae (79) and (81) each relate quantifications with the same range. It is in fact not uncommon that, all through a longer calculation, a given dummy has the same range; for brevity's sake, such constant ranges are stated once and not repeated over and over again. This opportunity for abbreviation was in fact one of the reasons for introducing the notion of the range in the first place.

Formula (81) has a partner, in hints referred to as "splitting the range":

(82) $[(\forall x:\ r.x:\ f.x) \wedge (\forall x:\ s.x:\ f.x) \equiv (\forall x:\ r.x \vee s.x:\ f.x)] \qquad .$

Proof We observe for any r , s , f

$$(\forall x:\ r.x:\ f.x) \wedge (\forall x:\ s.x:\ f.x)$$
$=$ {trading, 4 times}
$$(\forall x:\ \neg f.x:\ \neg r.x) \wedge (\forall x:\ \neg f.x:\ \neg s.x)$$
$=$ { $\forall$ distributes over $\wedge$ }
$$(\forall x:\ \neg f.x:\ \neg r.x \wedge \neg s.x)$$
$=$ {de Morgan}
$$(\forall x:\ \neg f.x:\ \neg(r.x \vee s.x))$$
$=$ {trading, twice}
$$(\forall x:\ r.x \vee s.x:\ f.x) \qquad .$$

(End of Proof.)

Another manifestation of associativity and symmetry is the postulate —referred to by "interchange of quantifications"— that for any f

(83) $[(\forall x::\ (\forall y::\ f.x.y)) \equiv (\forall y::\ (\forall x::\ f.x.y))] \qquad ,$

which invites us to observe for any r , s , f

$(\forall x: r.x: (\forall y: s.y: f.x.y))$

$=$ {trading, twice}

$(\forall x:: \neg r.x \lor (\forall y:: \neg s.y \lor f.x.y))$

$=$ { $\lor$ distributes over $\forall$ }

$(\forall x:: (\forall y:: \neg r.x \lor \neg s.y \lor f.x.y))$

$=$ {interchange of quantifications}

$(\forall y:: (\forall x:: \neg r.x \lor \neg s.y \lor f.x.y))$

$=$ { $\lor$ distributes over $\forall$ }

$(\forall y:: \neg s.y \lor (\forall x:: \neg r.x \lor f.x.y))$

$=$ {trading, twice}

$(\forall y: s.y: (\forall x: r.x: f.x.y))$.

Hence we have the more general

(84) $[(\forall x: r.x: (\forall y: s.y: f.x.y)) \equiv (\forall y: s.y: (\forall x: r.x: f.x.y))]$.

Also (84) will be referred to by "interchange of quantifications". Notice that in (84) each dummy carries with it its own range; as a result, no confusion arises when the ranges are left unmentioned.

Often we don't care which is the outer and which is the inner quantification. We cater to that by admitting more dummies after the $\forall$. By definition

(85) $[(\forall x,y:: f.x.y) \equiv (\forall x:: (\forall y:: f.x.y))]$.

The reader is invited to verify that it admits of the analogous generalization

(86) $[(\forall x,y: r.x \land s.x.y: f.x.y) \equiv$
            $(\forall x: r.x: (\forall y: s.x.y: f.x.y))]$.

In hints we refer to these rules as "nesting" or "unnesting".

Two rules confirm the status of the "everywhere" operator as universal quantifier. The first one is the analogue of (78), the distribution of $\lor$ over $\forall$: for any boolean structure X and any boolean scalar bs

(87) $bs \lor [X] \equiv [bs \lor X]$,

which is easily verified since *true* and *false* are the only boolean scalars.

As a special consequence of (87) we mention that the "everywhere" operator is strengthening, i.e.,

$[[X] \Rightarrow X]$ for all X .

Indeed, for any X

$$[[X] \Rightarrow X]$$
$$= \quad \{(67) \text{ with } X, Y := [X], X \}$$
$$[\neg[X] \vee X]$$
$$= \quad \{(87) \text{ with } bs := \neg[X] \}$$
$$\neg[X] \vee [X]$$
$$= \quad \{\text{Excluded Middle}\}$$
$$true \qquad .$$

The other rule, analogous to the interchange of universal quantifications, is the postulate

$$(88) \qquad [(\forall x: [r.x]: f.x)] \equiv (\forall x: [r.x]: [f.x]) \qquad ;$$

notice that in this case the range has to be a boolean scalar. So much for the status of the "everywhere" operator.

We now return to (79), the distribution of $\vee$ over $\forall$. With $X := true$, it yields

$$(89) \qquad [true \equiv (\forall x: r.x: true)] \qquad ;$$

from that, with the substitution $r := \neg f$ and trading, we derive

$$(90) \qquad [true \equiv (\forall x: false: f.x)] \qquad .$$

Because the equation

$$x: false$$

has no solutions —i.e., its solution set is empty— (90) is summarized as "with empty range, the universal quantification yields $true$". (Note how strongly (82) with $s.x := false$ suggests that $(\forall x: false: f.x)$ be the identity element of conjunction.)

The last two formulae tell us that some universal quantifications yield the value $true$; what is still lacking is a postulate telling us that some universal quantifications yield the value $false$: all postulates given so far would be satisfied if each universal quantification were equivalent to $true$! This is remedied by the postulate known as the "one-point rule", viz., that for any y of the same type as the dummy and any boolean function f on that type

$$(91) \qquad [(\forall x: [x = y]: f.x) \equiv f.y] \qquad .$$

Substituting in the above for f the constant function $false$, we get for any y of the same type as the dummy

$$(92) \qquad (\forall x: [x = y]: false) \equiv false \qquad ,$$

i.e., here we have a universal quantification equivalent to $\neg true$.

Remark An alternative would have been to postulate (92) and then to derive (91) as a theorem. It has the charm of a simpler postulate; the price to be paid is a longer proof. The reader may try to derive the one-point rule, given (92). We suggest the use of Leibniz's Rule. (*End of Remark.*)

Theorem For any set W

$$(93) \qquad (\forall x: \ x \in W: \ false) \equiv W = \varnothing \qquad .$$

Proof By case analysis, $W = \varnothing$ and $W \neq \varnothing$.

(i) $W = \varnothing$ In this case we borrow from set theory that the membership function of the empty set is the constant function *false* and observe

$(\forall x: \ x \in W: \ false)$
$= \quad \{ \ W = \varnothing \ \}$
$\quad (\forall x: \ x \in \varnothing: \ false)$
$= \quad \{$import from set theory$\}$
$\quad (\forall x: \ false: \ false)$
$= \quad \{(90)$ with $f.x := false \ \}$
$\quad true$
$= \quad \{ \ W = \varnothing \ \}$
$\quad W = \varnothing \qquad .$

(ii) $W \neq \varnothing$ In this case we borrow from set theory that from a non-empty set we may select an element and call it y . So, let $y \in W$; we then observe

$(\forall x: \ x \in W: \ false)$
$= \quad \{$Law of Absorption$\}$
$\quad (\forall x: \ x \in W \ \vee \ ([x = y] \ \wedge \ x \in W): \ false)$
$= \quad \{$Leibniz's Rule$\}$
$\quad (\forall x: \ x \in W \ \vee \ ([x = y] \ \wedge \ y \in W): \ false)$
$= \quad \{ \ y \in W \ \}$
$\quad (\forall x: \ x \in W \ \vee \ [x = y]: \ false)$
$= \quad \{$splitting the range$\}$
$\quad (\forall x: \ x \in W: \ false) \ \wedge \ (\forall x: \ [x = y]: \ false)$
$= \quad \{(92)\}$
$\quad (\forall x: \ x \in W: \ false) \ \wedge \ false$
$= \quad \{$pred. calc.$\}$
$\quad false$
$= \quad \{ \ W \neq \varnothing \ \}$
$\quad W = \varnothing \qquad .$

(End of Proof.)

Remark We included the first two steps of the proof (ii) above and did not start with

$(\forall x:\ x \in W:\ false)$
$=\ \ \{\ y \in W\ \}$
$\ \ (\forall x:\ x \in W\ \vee\ [x = y]:\ false)$

though, with their knowledge of set theory, most readers would not hesitate that "under the assumption $y \in W$", the two ranges are equivalent. The point is that this transformation has nothing to do with set theory: the formula

(94) $[r.y \Rightarrow (r.x\ \equiv\ r.x\ \vee\ [x = y])]$

is —as the reader may verify— perfectly general.

Furthermore, we would like to stress that, once the assumption $y \in W$ has been made, the first steps of proof (ii) don't come out of the blue at all.

Firstly, our calculation has to connect $y \in W$ with $x \in W$ from the demonstrandum; the one and only characteristic of functions —and the membership function is no exception— being that their application is equality-preserving, the boolean scalar $[x = y]$ has to enter the picture.

Secondly, once that need has been recognized, the Laws of Absorption are practically our only choice: they are the only formulae that express X in terms of X and a totally new Y and are thus also Laws of Introduction. The choice between the two is finally dictated by the already established need to apply the Rule of Leibniz. (*End of Remark.*)

And now we are ready for the next two theorems.

Theorem For any set W and boolean structure X

(95) $[(\forall x:\ x \in W:\ X)\ \equiv\ X\ \vee\ W = \varnothing]$.

Proof We observe for any X , W

$(\forall x:\ x \in W:\ X)$
$=\ \ \{\text{predicate calculus}\}$
$\ \ (\forall x:\ x \in W:\ X\ \vee\ false)$
$=\ \ \{\ \vee\ \text{ distributes over }\ \forall\ \}$
$\ \ X\ \vee\ (\forall x:\ x \in W:\ false)$
$=\ \ \{(93)\}$
$\ \ X\ \vee\ W = \varnothing$.

(*End of Proof.*)

Theorem For any set W , boolean structure X , and any function f from the (type of the) elements of W to boolean structures

$$(96)\quad W \neq \varnothing \Rightarrow [(\forall x\colon x\in W\colon X \wedge f.x) \equiv X \wedge (\forall x\colon x\in W\colon f.x)]\quad,$$

i.e., *provided the range is non-empty*, conjunction distributes over universal quantification.

Proof We observe for any non-empty W , any X , f

$\quad(\forall x\colon x\in W\colon X \wedge f.x)$
$=\quad\{\ \forall\ \text{distributes over}\ \wedge\ \}$
$\quad(\forall x\colon x\in W\colon X) \wedge (\forall x\colon x\in W\colon f.x)$
$=\quad\{(95)\}$
$\quad(X \vee W = \varnothing) \wedge (\forall x\colon x\in W\colon f.x)$
$=\quad\{\ W \neq \varnothing\ ,\ \text{identity element of}\ \vee\ \}$
$\quad X \wedge (\forall x\colon x\in W\colon f.x)\qquad.$

$\hfill$ *(End of Proof.)*

Thus, conjunction distributes over universal quantification only if the range is non-empty, whereas the distribution of disjunction over universal quantification is unrestricted. In what follows, we shall regularly encounter consequences of this difference.

Next we observe for any y (of the type of the dummy) and any f (of the appropriate type)

$\quad(\forall x\colon\colon f.x) \wedge f.y$
$=\quad\{\text{one-point rule}\}$
$\quad(\forall x\colon\colon f.x) \wedge (\forall x\colon [x = y]\colon f.x)$
$=\quad\{\text{splitting the range}\}$
$\quad(\forall x\colon \mathit{true} \vee [x = y]\colon f.x)$
$=\quad\{\text{zero element of}\ \vee\ \}$
$\quad(\forall x\colon\colon f.x)\qquad.$

Hence the Rule of Instantiation: for any y , f

$$(97)\qquad [(\forall x\colon\colon f.x) \Rightarrow f.y]$$

and, with $f.z := [f.z]$, for any y

$$(98)\qquad (\forall x\colon\colon [f.x]) \Rightarrow [f.y]\qquad.$$

It is this formula that allows us to identify the phrase

$\qquad$ "We have, for any x , $[f.x]$."

with

$\qquad$ "We have $(\forall x\colon\colon [f.x])$." $\qquad.$

Both allow us to conclude, for any y , that we have $[f.y]$; from an operational point of view, the two phrases are equivalent. This explains why "∀" is traditionally read as "for all" and is called "the universal quantifier". It is customary to use the first phrasing in verbal contexts, but to use the explicit universal quantification as soon as the expression is needed as subexpression in a formal context. We have followed this custom (and shall continue to do so). The custom is nicely illustrated by (97), in which the universal quantification over x is indicated explicitly, and for which we did not feel compelled to write

$$(\forall f, y :: \ [(\forall x :: \ f.x) \Rightarrow f.y])$$,

although it would have been perfectly correct.

Now the time has come to take a second look at some of our conventions.

To begin with, we allowed ourselves complete freedom in the choice of our variables, and when expressing, for instance, that negation is its own inverse, it did not matter whether we wrote

$$[\neg\neg X \equiv X] \qquad \text{or} \qquad [\neg\neg Y \equiv Y] \qquad .$$

We can now express that they are short for the universal quantifications

$$(\forall X :: \ [\neg\neg X \equiv X]) \qquad \text{and} \qquad (\forall Y :: \ [\neg\neg Y \equiv Y]) \qquad ,$$

respectively, whose equivalence is a special case of

(99) $[(\forall x :: \ f.x) \equiv (\forall y :: \ f.y)]$.

The aforementioned freedom in the choice of our variables is the same freedom we have in the choice of dummies: those variables are dummies. So far, so good. But (99) reveals a problem we have glossed over. It stresses that $(\forall x :: \ f.x)$ is *not* a function of x , and this raises two questions:

(i) is $(\forall x :: \ f.x)$ a function, and, if so, of what?
(ii) to what extent need we depart from our general rule that expressions are functions of their subexpressions?

Question (i) is answered by postulating that $(\forall x :: \ f.x)$ is a function of f , i.e., that we have Leibniz's Rule

$$[f \equiv g] \Rightarrow [(\forall x :: \ f.x) \equiv (\forall x :: \ g.x)] \qquad ,$$

which —see (3,4)— we rewrite as

(100) $[(\forall x :: \ f.x \equiv g.x)] \Rightarrow [(\forall x :: \ f.x) \equiv (\forall x :: \ g.x)]$.

Question (ii) can be answered in two ways. It deals with the status of "$f.x$" in $(\forall x :: \ f.x)$: syntactically "$f.x$" is a subexpression of $(\forall x :: \ f.x)$, but is the latter really a function of that subexpression? The one answer says: no, not really. We could (and perhaps should) have chosen a new notation in which the dummy has been eliminated, say $(\forall:: f)$ instead of $(\forall x:: f.x)$. The new notation would stress that $(\forall:: f)$ is a function of its subexpression f . Following that line of thought, all our formulae that equate expressions should be rephrased so as to equate functions; for instance, instead of $[\neg\neg X \equiv X]$ we might write $[\neg \circ \neg \equiv I]$, indicating that negation, functionally composed with itself, yields the identity function. But the price gets heavy. It is, for instance, quite possible to state de Morgan's Law

$$[\neg(X \vee Y) \equiv \neg X \wedge \neg Y]$$

as the equality of two functions, each defined on a pair of boolean structures. In one of our proofs we used the simple instantiation X , $Y := r.x$, $s.x$

$$[\neg(r.x \vee s.x) \equiv \neg r.x \wedge \neg s.x] \qquad ,$$

which expresses the equality of two functions of x . Such instantiations become sizeable exercises in functional abstraction, composition, and application. The other answer is to accept $f.x$ as a subexpression of $(\forall x:: f.x)$ and to consider (100) with the antecedent universally quantified over the dummy as the proper generalization of the Rule of Leibniz for subexpressions, some of whose global variables are dummies. We have adopted the latter approach.

The hint "Leibniz" will also be used to refer to (100). By the time its use has become second nature, the hint will be suppressed. We give the format of some typical proof steps:

$\qquad [(\forall x:: f.x) \equiv (\forall x:: g.x)]$
$\Leftarrow \quad \{\text{Leibniz}\}$
$\qquad [(\forall x:: f.x \equiv g.x)] \qquad ,$

$\qquad [(\forall x:: f.x) \equiv (\forall x:: g.x)]$
$= \quad \{\text{Leibniz; since } [f.x \equiv g.x] \}$
$\qquad true \qquad ,$

$\qquad (\forall x:: f.x)$
$= \quad \{\text{Leibniz; since } [f.x \equiv g.x] \}$
$\qquad (\forall x:: g.x) \qquad .$

We can now prove the stronger "punctual Leibniz"; for the sake of generality we introduce a range at the same time.

Theorem "$\forall$ is punctual", i.e., for any r , f , g

(101) $\qquad [(\forall x: r.x: f.x \equiv g.x) \Rightarrow$
$\qquad\qquad ((\forall x: r.x: f.x) \equiv (\forall x: r.x: g.x))] \qquad .$

Proof We observe for any r , f , g

(101)
$=$ $\{(51)\}$
$[(\forall x:\ r.x:\ f.x \equiv g.x) \wedge (\forall x:\ r.x:\ f.x) \equiv$
$(\forall x:\ r.x:\ f.x \equiv g.x) \wedge (\forall x:\ r.x:\ g.x)]$
$=$ $\{\ \forall\ \text{distributes over}\ \wedge\ \}$
$[(\forall x:\ r.x:\ (f.x \equiv g.x) \wedge f.x) \equiv (\forall x:\ r.x:\ (f.x \equiv g.x) \wedge g.x)]$.

Proceeding with the left-hand side we observe

$(\forall x:\ r.x:\ (f.x \equiv g.x) \wedge f.x)$
$=$ $\{(29)\ \text{with}\ X, Y := f.x, g.x\ \text{yields}\ [(f.x \equiv g.x) \wedge f.x \equiv f.x \wedge g.x]\ \}$
$(\forall x:\ r.x:\ f.x \wedge g.x)$.

Since the right-hand side yields the same expression, this concludes our proof.

(End of Proof.)

Since "everywhere" is monotonic, we conclude from (101) that the nonpunctual Leibniz (100) also holds for a range differing from *true* .

And now we should be ready to show that universal quantification is monotonic with respect to implication, and even punctually so.

Theorem For any r , f , g

(102) $[(\forall x:\ r.x:\ f.x \Rightarrow g.x) \Rightarrow ((\forall x:\ r.x:\ f.x) \Rightarrow (\forall x:\ r.x:\ g.x))]$.

Proof We observe for any r , f , g

$(\forall x:\ r.x:\ f.x) \Rightarrow (\forall x:\ r.x:\ g.x)$
$=$ $\{\text{pred. calc.}\}$
$(\forall x:\ r.x:\ f.x) \wedge (\forall x:\ r.x:\ g.x) \equiv (\forall x:\ r.x:\ f.x)$
$=$ $\{\ \forall\ \text{distributes over}\ \wedge\ \}$
$(\forall x:\ r.x:\ f.x \wedge g.x) \equiv (\forall x:\ r.x:\ f.x)$
$\Leftarrow$ $\{\ \forall\ \text{is punctual}\}$
$(\forall x:\ r.x:\ f.x \wedge g.x \equiv f.x)$
$=$ $\{\text{pred. calc.}\}$
$(\forall x:\ r.x:\ f.x \Rightarrow g.x)$.

(End of Proof.)

We give two further theorems.

Theorem We have for any r, f, g, h of proper types

(103) $[(\forall x:\ r.x:\ [f.x = g.x]) \Rightarrow (\forall x:\ r.x:\ h.(f.x) = h.(g.x))]$.

Proof We observe for any r, f, g, h

  (103)
$\Leftarrow$ {monotonicity of $\forall$ }
  $[(\forall x:\ r.x:\ [f.x = g.x] \Rightarrow h.(f.x) = h.(g.x))]$
$=$ {Leibniz}
  $[(\forall x:\ r.x:\ true)]$
$=$ {(89)}
  true   .

                                                                               (*End of Proof.*)

Theorem Wc have for any r, f, g, h of the appropriate types

(104) $[(\forall x:\ r.x:\ [f.x = g.x]) \Rightarrow$
                    $((\forall x:\ r.x:\ h.(f.x)) \equiv (\forall x:\ r.x:\ h.(g.x)))]$.

Proof We observe for any r, f, g, h

  $(\forall x:\ r.x:\ h.(f.x)) \equiv (\forall x:\ r.x:\ h.(g.x))$
$\Leftarrow$ { $\forall$ is punctual}
  $(\forall x:\ r.x:\ h.(f.x) \equiv h.(g.x))$
$\Leftarrow$ {(103)}
  $(\forall x:\ r.x:\ [f.x = g.x])$.

                                                                               (*End of Proof.*)

An important manipulation involving an invertible function is referred to as "transforming the dummy". In the special case that the invertible function is the identity function, the usual hint is "renaming the dummy". It relies on the next theorem.

Theorem We have for any r, f, and invertible t

(105) $[(\forall x:\ r.x:\ f.x) \equiv (\forall y:\ r.(t.y):\ f.(t.y))]$.

Proof Using trading we see that it suffices to prove the theorem with the range *true* . For that range we shall establish the result by mutual implication, i.e., our demonstranda are

(106) $[(\forall x::\ f.x) \Rightarrow (\forall y::\ f.(t.y))]$ and

(107) $[(\forall x::\ f.x) \Leftarrow (\forall y::\ f.(t.y))]$.

Proof of (106) We observe for any f and —not necessarily invertible— t

$[(\forall x:: \ f.x) \Rightarrow (\forall y:: \ f.(t.y))]$
= $\quad \{$ "$X \Rightarrow$" distributes over $\forall$ like "$\neg X \vee$" does$\}$
$[(\forall y:: \ (\forall x:: \ f.x) \Rightarrow f.(t.y))]$
= $\quad \{(97)$ with $y := t.y \ \}$
$[(\forall y:: \ true)]$
= $\quad \{(89)\}$
$\quad true$.

(End of Proof of (106).)

Proof of (107) We observe for any f and invertible t

$(\forall y:: \ f.(t.y))$
= $\quad \{$definition of functional composition$\}$
$(\forall y:: \ (f \circ t).y)$
$\Rightarrow \quad \{(106)$ with $x, y, f, t := y, x, f \circ t, t^{-1} \ \}$
$(\forall x:: \ (f \circ t).(t^{-1}.x))$
= $\quad \{$definition of functional composition$\}$
$(\forall x:: \ f.((t \circ t^{-1}).x))$
= $\quad \{ \ t \circ t^{-1}$ is the identity function$\}$
$(\forall x:: \ f.x)$.

(End of Proof of (107).)
(End of Proof.)

Remark Firstly, we would like to point out that the above argument is beautifully disentangled. The theorem is about an invertible t , but the first subproof holds for any t . Had we approached the problem saying to ourselves: "Let us try to use our data one at a time, so let us investigate what we can derive without using that t is invertible.", we would probably have concluded that for this theorem a ping-pong argument is appropriate. Note also that —via (97)— only the first subproof refers to earlier properties of universal quantification. Secondly, we would like to draw the reader's attention to the second subproof: without a proper notation for functional composition it would have been hard to describe explicitly how the latter's associativity is being exploited. It illustrates once more the importance of adequate notation. *(End of Remark.)*

Later in this booklet we shall occupy ourselves extensively with the question of whether function applications distribute over universal quantification, i.e., whether we have

$$[f.(\forall X: \ X \in W: \ X) \equiv (\forall X: \ X \in W: \ f.X)]$$.

Because our functions tend to be monotonic, we shall frequently appeal to the next

Theorem For any monotonic function f (from boolean structures to boolean structures) and any bag W (of boolean structures)

(108) $[f.(\forall X: X \in W: X) \Rightarrow (\forall X: X \in W: f.X)]$.

Proof We observe for any monotonic f and any W

(108)
$=$ {renaming a dummy}
    $[f.(\forall X: X \in W: X) \Rightarrow (\forall Y: Y \in W: f.Y)]$
$=$ { "$Z \Rightarrow$" distributes over $\forall$ }
    $[(\forall Y: Y \in W: f.(\forall X: X \in W: X) \Rightarrow f.Y)]$
$=$ { $Y \in W$ is a boolean scalar, interchange}
    $(\forall Y: Y \in W: [f.(\forall X: X \in W: X) \Rightarrow f.Y])$
$\Leftarrow$ { f is monotonic, and so is $\forall$ }
    $(\forall Y: Y \in W: [(\forall X: X \in W: X) \Rightarrow Y])$
$=$ { $Y \in W$ is a boolean scalar, interchange}
    $[(\forall Y: Y \in W: (\forall X: X \in W: X) \Rightarrow Y)]$
$=$ {(109), see below}
    true   .

Note Formula (97) can be extended to include a range. We have indeed

(109) $[(\forall y: r.y: (\forall x: r.x: f.x) \Rightarrow f.y)]$,

as is established by observing for any r , f

(109)
$=$ {trading, twice}
    $[(\forall y:: \neg r.y \vee ((\forall x:: \neg r.x \vee f.x) \Rightarrow f.y))]$
$=$ { $[X \vee (Y \Rightarrow Z) \equiv Y \Rightarrow X \vee Z]$ }
    $[(\forall y:: (\forall x:: \neg r.x \vee f.x) \Rightarrow \neg r.y \vee f.y)]$
$=$ {(97) and predicate calculus}
    true   .

Alternatively, (109) is phrased as: for all y

(110) $[(\forall x: r.x: f.x) \wedge r.y \Rightarrow f.y]$.

                                                            (*End of Note.*)

When we use the term "monotonic", we usually mean "monotonic with respect to implication", i.e., f is monotonic means for a function from boolean structures to boolean structures

$$[X \Rightarrow Y] \Rightarrow [f.X \Rightarrow f.Y]$$.

From "preserving some order" the notion of monotonicity has been generalized to the transfer of some order in the argument into some order of the function value. In particular, for a function f from the natural numbers to boolean structures, "being monotonic" means that we have

(111) $(\forall i,j:\ 0 \leqslant i \leqslant j:\ [f.i \Rightarrow f.j])$ or

(112) $(\forall i,j:\ 0 \leqslant i \leqslant j:\ [f.i \Leftarrow f.j])$.

If we have to distinguish, we say for (111) "f is weakening" and for (112) "f is strengthening". For weakening and strengthening functions we have the following two theorems.

Theorem   For a weakening boolean function f on the natural numbers

(113) $[(\forall i:\ 0 \leqslant i:\ f.i) \equiv f.0]$ and

(114) $n > 0 \Rightarrow [(\forall i:\ 0 \leqslant i < n:\ f.i) \equiv f.0]$.

Proof   We observe for any weakening f

$\quad (\forall i:\ 0 \leqslant i:\ f.i)$
$=\quad \{(110),\ \text{with}\ x, y, r.x := i, 0, 0 \leqslant i\ \text{and}\ 0 \leqslant 0,\ \text{and}\ (46)\}$
$\quad f.0 \wedge (\forall i:\ 0 \leqslant i:\ f.i)$
$=\quad \{(96),\ \text{range is non-empty}\}$
$\quad (\forall i:\ 0 \leqslant i:\ f.0 \wedge f.i)$
$=\quad \{\ f\ \text{is weakening, (111) with}\ i, j := 0, i\ \text{yields}$
$\qquad\qquad (\forall i:\ 0 \leqslant i:\ [f.0 \wedge f.i \equiv f.0])\ \}$
$\quad (\forall i:\ 0 \leqslant i:\ f.0)$
$=\quad \{\text{range is non-empty}\}$
$\quad f.0 \qquad .$

The proof of (114) is left to the reader.

(End of Proof.)

Corollary   For strengthening f and natural n

(115) $[(\forall i:\ 0 \leqslant i \leqslant n:\ f.i) \equiv f.n]$.

Proof   From (114) by transforming the dummy.

(End of Proof.)

Theorem   For a boolean function f of two arguments that is weakening in both arguments or is strengthening in both arguments

(116) $[(\forall i,j:\ 0 \leqslant i \wedge 0 \leqslant j:\ f.i.j) \equiv (\forall i:\ 0 \leqslant i:\ f.i.i)]$.

(This theorem can be generalized to functions of more arguments.)

Proof We give separate proofs for weakening and strengthening f .

For any weakening f we observe first

$(\forall i,j:\ 0 \leqslant i \leqslant j:\ [f.i.i \Rightarrow f.j.j])$
$\Leftarrow$ {transitivity of $\Rightarrow$ }
$\quad (\forall i,j:\ 0 \leqslant i \leqslant j:\ [f.i.i \Rightarrow f.j.i] \wedge [f.j.i \Rightarrow f.j.j])$
$=$ { f weakening in 1st and in 2nd argument}
true   ,

hence $f.i.i$ is a weakening function of i . Furthermore, we observe

$(\forall i,j:\ 0 \leqslant i \wedge 0 \leqslant j:\ f.i.j)$
$=$ {nesting}
$\quad (\forall i:\ 0 \leqslant i:\ (\forall j:\ 0 \leqslant j:\ f.i.j))$
$=$ {(113); $f.i.j$ is a weakening function of j }
$\quad (\forall i:\ 0 \leqslant i:\ f.i.0)$
$=$ {(113); $f.i.0$ is a weakening function of i }
$\quad f.0.0$
$=$ {(113); $f.i.i.$ is a weakening function of i }
$\quad (\forall i:\ 0 \leqslant i:\ f.i.i)$.

So much for weakening f . For strengthening f we observe first

$(\forall i,j:\ 0 \leqslant i \wedge 0 \leqslant j:\ f.i.j)$
$=$ {predicate calculus and transitivity of $\leqslant$ }
$\quad (\forall i,j:\ 0 \leqslant i \leqslant j \vee 0 \leqslant j \leqslant i:\ f.i.j)$
$=$ {splitting the range}
$\quad (\forall i,j:\ 0 \leqslant i \leqslant j:\ f.i.j) \wedge (\forall i,j:\ 0 \leqslant j \leqslant i:\ f.i.j)$.

Focussing our attention on the left-hand conjunct we observe

$(\forall i,j:\ 0 \leqslant i \leqslant j:\ f.i.j)$
$=$ {predicate calculus and transitivity of $\leqslant$ }
$\quad (\forall j,i:\ 0 \leqslant j \wedge 0 \leqslant i \leqslant j:\ f.i.j)$
$=$ {nesting}
$\quad (\forall j:\ 0 \leqslant j:\ (\forall i:\ 0 \leqslant i \leqslant j:\ f.i.j))$
$=$ {(115); $f.i.j$ is a strengthening function of i }
$\quad (\forall j:\ 0 \leqslant j:\ f.j.j)$.

For reasons of symmetry the other conjunct has the same value. With the idempotence of the conjunction, these two observations settle the matter in the case of strengthening f .

(End of Proof.)

When introducing universal quantification, we said that it was a generalization of conjunction. This was inspired by the following observation:

$$(\forall i: \ 0 \leqslant i < n + 1: \ f.i)$$
$=$ {arithmetic}
$$(\forall i: \ 0 \leqslant i < n \ \lor \ i = n: \ f.i)$$
$=$ {splitting the range and one-point rule}
$$(\forall i: \ 0 \leqslant i < n: \ f.i) \ \land \ f.n \qquad ,$$

recursive application of which tempts some to write

$$[(\forall i: \ 0 \leqslant i < n + 1: \ f.i) \ \equiv \ f.0 \ \land \ f.1 \ \land \ \cdots \ \land \ f.n]$$

or the even worse

$$[(\forall i: \ 0 \leqslant i: \ f.i) \ \equiv \ f.0 \ \land \ f.1 \ \land \ \cdots] \qquad .$$

As the reader will have noticed, we hardly stressed this connection between universal quantification and conjunction. It is not only that we dislike the ominous three dots and that it only works for finite ranges, though —thanks to trading— the range is less a property of the quantified expression and more of the way it has been written down. Our greatest objection is that the analogy sneakily suggests that quantification only provides an (in this case badly needed!) shorthand for "infinite expressions". But that metaphor evokes more questions than it answers, and, even worse, questions that we propose to deal with by not raising them, since we don't seem to need the answers. We prefer to view $(\forall x:: \ f.x)$ as a very finite expression in the one variable f .

So much for universal quantification.

*   *   *

To the extent that universal quantification is the analogue of conjunction, existential quantification is the analogue of disjunction. It can now be defined by the analogue of de Morgan's Law

(117) $[(\exists x: \ r.x: \ f.x) \equiv \neg(\forall x: \ r.x: \ \neg f.x)]$.

It has all the properties one would expect it to have. For the sake of completeness we give a number of them. They are the duals of the corresponding properties of universal quantification, and their verification is left to the reader.

• "trading" —see (77)—

$$[(\exists x: \ r.x: \ f.x) \ \equiv \ (\exists x:: \ r.x \land f.x)] \qquad ,$$

from which we see that existential quantification is symmetric in range and term.

- " $\wedge$ distributes over $\exists$ " —see (79)—

$$[X \wedge (\exists x: \; r.x: \; f.x) \equiv (\exists x: \; r.x: \; X \wedge f.x)] \qquad .$$

- " $\exists$ distributes over $\vee$ " —see (81)—

$$[(\exists x: \; r.x: \; f.x) \vee (\exists x: \; r.x: \; g.x) \equiv (\exists x: \; r.x: \; f.x \vee g.x)] \qquad .$$

- "splitting the range" —see (82)—

$$[(\exists x: \; r.x: \; f.x) \vee (\exists x: \; s.x: \; f.x) \equiv (\exists x: \; r.x \vee s.x: \; f.x)] \qquad .$$

(In view of the symmetry in range and term, the last two formulae are not very different.)

- "interchange of existential quantifications" is as before —see (84)— ; note that in "nesting" and "unnesting" —see (86)— the conjunction in the range is maintained:

$$[(\exists x,y: \; r.x \wedge s.x.y: \; f.x.y) \equiv (\exists x: \; r.x: \; (\exists y: \; s.x.y: \; f.x.y)] \qquad .$$

- $[false \equiv (\exists x: \; r.x: \; false)]$ —see (89)—
  $[false \equiv (\exists x: \; false: \; f.x)]$ —see (90)—   ,

i.e., existential quantification over the empty range yields *false* .

- "one-point rule" —see (91)—

$$[(\exists x: \; [x = y]: \; f.x) \equiv f.y] \qquad .$$

(The one-point rule holds for the quantifying generalizations of all associative operators, sum, product, maximum, minimum, etc. See also Dirac's delta function

$$\int \delta.(x - y) f.x \; dx = f.y \qquad .)$$

-     $(\exists x:: \; x \in W) \equiv W \neq \varnothing$ —see (93)—
      $[(\exists x: \; x \in W: \; X) \equiv X \wedge W \neq \varnothing]$ —see (95)—   .

-     $[(\exists x:: \; f.x) \Leftarrow f.y]$ —see (97)—   .

- "punctual generalized Leibniz" —see (101)—

$$[(\forall x: \; r.x: \; f.x \equiv g.x) \Rightarrow ((\exists x: \; r.x: \; f.x) \equiv (\exists x: \; r.x: \; g.x))] \qquad .$$

- " $\exists$ is punctually monotonic" —see (102)—

$$[(\forall x: \; r.x: \; f.x \Rightarrow g.x) \Rightarrow ((\exists x: \; r.x: \; f.x) \Rightarrow (\exists x: \; r.x: \; g.x))] \qquad .$$

- "transforming the dummy", —see (105)— i.e., for invertible t

$$[(\exists x: \; r.x: \; f.x) \equiv (\exists y: \; r.(t.y): \; f.(t.y))] \qquad .$$

- We have for monotonic f —see (108)—

$$[f.(\exists X: \; X \in W: \; X) \Leftarrow (\exists X: \; X \in W: \; f.X)] \qquad .$$

(Note that, just as in the existential form of instantiation, the implication is here the other way round.)

● For f a strengthening boolean function on the natural numbers —see (113)—

$$[(\exists i: \ 0 \leqslant i: \ f.i) \equiv f.0] \qquad .$$

● For a boolean function of two arguments that is weakening in both arguments or strengthening in both arguments —see (116)—

$$[(\exists i, j: \ 0 \leqslant i \wedge 0 \leqslant j: \ f.i.j) \equiv (\exists i: \ 0 \leqslant i: \ f.i.i)] \qquad .$$

So much for the existential transcription of former results. There are furthermore two (ugly) formulae involving the two quantifications and the implication:

(118) $[(\exists x: \ r.x: \ f.x) \Rightarrow Y \equiv (\forall x: \ r.x: \ f.x \Rightarrow Y)]$

and

(119) $[(\exists x:: \ r.x) \Rightarrow ((\forall x: \ r.x: \ f.x) \Rightarrow Y \equiv (\exists x: \ r.x: \ f.x \Rightarrow Y))]$,

in which the antecedent $(\exists x:: \ r.x)$ corresponds to the proviso that the range be non-empty. We leave their verification to the reader.

$$* \qquad * \qquad *$$

In Chap. 1 we mentioned that a pair —in general, an n-tuple— of structures may be considered as a single structure on a doubled —in general, an n-fold— space. The "everywhere" operators corresponding to those different spaces are connected by

(120) $[(X, Y)] \equiv [X] \wedge [Y] \qquad .$

(We give the formulae for pairs, leaving the generalization to n-tuples to the reader.)

The punctuality of our operators is reflected by the fact that they distribute over pair-forming, e.g.,

$$[\neg(X, \ Y) \equiv (\neg X, \ \neg Y)]$$
$$[(X0, \ Y0) \equiv (X1, \ Y1) \equiv (X0 \equiv X1, \ Y0 \equiv Y1)]$$
$$[(X0, \ Y0) \vee (X1, \ Y1) \equiv (X0 \vee X1, \ Y0 \vee Y1)]$$
$$[(\exists x:: \ (f.x, \ g.x)) \equiv ((\exists x:: \ f.x), (\exists x:: \ g.x))] \qquad ,$$

etc. (See also (6,47) and (6,48).)

$$* \qquad * \qquad *$$

In the remainder of this little monograph we deal almost exclusively with boolean structures. From here on we shall denote them by their usual name, "predicates"; functions from boolean structures to boolean structures will be called "predicate transformers".

CHAPTER 6

Some properties of predicate transformers

In this chapter we define and explore a number of properties that predicate transformers may or may not enjoy. It is a preparation for the later chapters in which we analyse in terms of these properties the predicate transformers that will be used to define programming language semantics. The purpose of that later analysis is to justify the procedures followed in proving properties of programs.

As said at the end of the preceding chapter, the boolean structures we encounter in programming are traditionally called "predicates" —the terms "conditions" and "assertions" are used also— and functions from predicates to predicates are traditionally called "predicate transformers". (This term was inspired by the circumstance that for the original —and still most common— predicate transformers f , the predicates X and $f.X$ were boolean structures on the same space —viz., the state space of the program under consideration— and that such an f was viewed as an operator "transforming" any given predicate X into the corresponding predicate $f.X$.)

A predicate transformer being a function from predicates to predicates is reflected in the availability of Leibniz's Rule, i.e., we have for any predicate transformer f and any predicates X and Y (on the appropriate space)

(0) $[X \equiv Y] \Rightarrow [f.X \equiv f.Y]$.

Note that predicate transformers need not be punctual functions; in fact the predicate transformers of interest have a strong tendency to be nonpunctual.

Remark   A formula (without quantification) in which the "everywhere" operator is applied to a punctual function of the variables occurring in it can be verified by substituting for the variables all possible combinations of *true* and *false* and subsequently simplifying the resulting expressions to the boolean scalar [*true*] . In —the first part of— Chap. 5, we chose not to do so and to use the algebraic style because the latter's applicability is not confined to punctual functions of the variables and our main interest is in nonpunctual predicate transformers. (*End of Remark.*)

As said, this chapter is about properties of predicate transformers. We have already encountered a property that a predicate transformer may or may not enjoy: monotonicity with respect to implication, or "monotonicity" for short. Formally, we have for any predicate transformer f

(1)   (f is monotonic) $\equiv$ $(\forall X, Y :: [X \Rightarrow Y] \Rightarrow [f.X \Rightarrow f.Y])$.

As we shall see shortly, monotonicity is the weakest of the properties to be introduced in this chapter. The additional properties come in pairs: each has its dual. For brevity's sake we introduce the duality first, because it saves us the trouble of giving a whole series of proofs in two versions.

Remark Monotonicity, as said, is the weakest of the properties considered in this chapter. It is, in fact, so weak that it is enjoyed by almost all predicate transformers of interest. We could mention it prior to the introduction of the duality because —as we shall see shortly— monotonicity is its own dual. (*End of Remark.*)

The conjugate of a predicate transformer

With f^* we denote the predicate transformer that is called "the conjugate" of predicate transformer f .

Apology We apologize for yet another notational convention: the postfix operator denoted by the raised star. Though the notion of the conjugate is of relevance through most of the rest of this little monograph, the notation with the raised star will hardly be used outside this chapter. The raised star has a higher binding power than functional application; for instance $\neg f^*.X$ should be parsed as $\neg((f^*).X)$. (*End of Apology.*)

Taking the conjugate is defined by the fact that we have for any predicate transformer f and any predicate X

(2) $[f^*.X \equiv \neg f.(\neg X)]$.

The term "conjugate" is justified by the circumstance that, if one predicate transformer is the conjugate of another, they are each other's conjugates, as follows from the following

Theorem Taking the conjugate is its own inverse, i.e., we have for any predicate transformer f and any predicate X

(3) $[f^{**}.X \equiv f.X]$.

Proof We observe for any f , X

$\quad f^{**}.X$
$= \quad \{(2) \text{ with } f := f^* \}$
$\quad \neg f^*.(\neg X)$
$= \quad \{(2) \text{ with } X := \neg X \}$
$\quad \neg\neg f.(\neg\neg X)$
$= \quad \{\text{double negation, twice}\}$
$\quad f.X \qquad .$

(*End of Proof.*)

Remark Note that some functions are their own conjugate, e.g., the identity function and the negation. (*End of Remark.*)

The properties to be introduced in a moment are paired by the circumstance that if f enjoys one property of a pair, f^* enjoys the other property of the pair. This is the duality we mentioned. That monotonicity could be introduced prior to the introduction of that duality is a consequence of the

Theorem For any predicate transformer f

(4)   (f is monotonic) $\equiv$ (f^* is monotonic)   .

Proof We observe for any f

$\quad (f^* \text{ is monotonic})$
$= \quad \{(1), \text{ i.e., def. of monotonicity}\}$
$\quad (\forall X,Y: [X \Rightarrow Y]: [f^*.X \Rightarrow f^*.Y])$
$= \quad \{(2), \text{ i.e., definition of conjugate}\}$
$\quad (\forall X,Y: [X \Rightarrow Y]: [\neg f.(\neg X) \Rightarrow \neg f.(\neg Y)])$
$= \quad \{\text{contra-positive, twice}\}$
$\quad (\forall X,Y: [\neg Y \Rightarrow \neg X]: [f.(\neg Y) \Rightarrow f.(\neg X)])$
$= \quad \{\text{transforming the dummies: negation is invertible}\}$
$\quad (\forall X,Y: [Y \Rightarrow X]: [f.Y \Rightarrow f.X])$
$= \quad \{\text{definition of monotonicity}\}$
$\quad (f \text{ is monotonic}) \qquad .$

(*End of Proof.*)

We now define the two central concepts that will occupy us for the remainder of this chapter: conjunctivity and disjunctivity. For any predicate transformer f and any bag V of predicates we define

(5) $(f$ is conjunctive over $V) \equiv$
       $[f.(\forall X:\ X \in V:\ X) \equiv (\forall X:\ X \in V:\ f.X)]$

(6) $(f$ is disjunctive over $V) \equiv$
       $[f.(\exists X:\ X \in V:\ X) \equiv (\exists X:\ X \in V:\ f.X)]$.

In words: the conjunctivity of f describes the extent to which application of f distributes over universal quantification, its disjunctivity describes how its application distributes over existential quantification.

For brevity's sake we introduce the notion of "the conjugate of a bag of predicates". For a bag V of predicates we obtain its conjugate V^* by negating all its predicates. Consequently we have —with $\in$ having a higher binding power than $\equiv$ —

(7) $X \in V^* \equiv (\neg X) \in V$

(8) $V^{**} = V$.

Now we can formulate the theorem that forms the basis for the duality alluded to.

Theorem We have for any predicate transformer f and any bag V of predicates

(9) $(f$ is conjunctive over $V) \equiv (f^*$ is disjunctive over $V^*)$.

Proof We observe for any f , V

   $(f^*$ is disjunctive over $V^*)$
= $\{(6),$ i.e., def. of disjunctivity$\}$
   $[f^*.(\exists X:\ X \in V^*:\ X) \equiv (\exists X:\ X \in V^*:\ f^*.X)]$
= $\{(2)$ and $(7),$ i.e., notions of conjugate$\}$
   $[\neg f.(\neg(\exists X:\ (\neg X) \in V:\ X)) \equiv (\exists X:\ (\neg X) \in V:\ \neg f.(\neg X))]$
= $\{$transforming the dummies: negation is invertible$\}$
   $[\neg f.(\neg(\exists X:\ X \in V:\ \neg X)) \equiv (\exists X:\ X \in V:\ \neg f.X)]$
= $\{$negating both sides$\}$
   $[f.(\neg(\exists X:\ X \in V:\ \neg X)) \equiv \neg(\exists X:\ X \in V:\ \neg f.X)]$
= $\{$de Morgan, twice$\}$
   $[f.(\forall X:\ X \in V:\ X) \equiv (\forall X:\ X \in V:\ f.X)]$
= $\{(5),$ i.e., def. of conjunctivity$\}$
   $(f$ is conjunctive over $V)$.

                                                              (*End of Proof.*)

The different types of junctivity

We use the term "junctivity" in sentences that are applicable to both conjunctivity and disjunctivity; the adjective "junctive" is used in a similar fashion.

In the preceding section we have met the notions of a predicate transformer f being junctive over a specific bag V of predicates. We look, however, for a notion of junctivity pertaining to f all by itself, i.e., a notion of junctivity that does not refer to a specific bag V. We know how to eliminate V: quantify over it! And so we arrive at the notion of "universal junctivity":

(10)   (f is universally junctive) $\equiv (\forall V:: (f$ is junctive over $V))$.

Remark There seems little point in pursuing the alternative of eliminating V by means of existential quantification because any f is junctive over any singleton bag V. (*End of Remark.*)

But universal junctivity is a strong property! By instantiating the right-hand side of (10) with $V := \varnothing$ we get

(f is universally junctive) $\Rightarrow (f$ is junctive over $\varnothing)$

with the immediate consequences

(11)   (f is universally conjunctive) $\Rightarrow [f.true \equiv true]$

(12)   (f is universally disjunctive) $\Rightarrow [f.false \equiv false]$.

We can get weaker junctivity properties by strengthening the range of V in the right-hand side of (10). For instance, in order to avoid the consequents of (11) and (12), we could constrain the universal quantification to all non-empty V. In a moment we shall, indeed, define a type of junctivity —weaker than universal junctivity— by just doing that.

But before introducing different types of junctivity, we should remember (9), the basis for the duality we are aiming at. We wish to introduce such junctivity types that

(13)   (the conjunctivity type of f) = (the disjunctivity type of f^*)

is ensured.

Introducing temporarily the notion of "r-junctivity", we have for some (scalar) boolean function r

(14)   (f is r-conjunctive) $\equiv$
            ($\forall V$: $r.V$: f is conjunctive over V)   and

(15)   (f is r-disjunctive) $\equiv$
            ($\forall V$: $r.V$: f is disjunctive over V)   .

In view of (13), r should be such that

(16)   (f is r-conjunctive) $\equiv$ (f^* is r-disjunctive)   .

In order to analyse this requirement we observe

   (f^* is r-disjunctive)
= {(15) with $f := f^*$ }
   ($\forall V$: $r.V$: f^* is disjunctive over V)
= {transforming the dummy: $*$ is invertible}
   ($\forall V$: $r.V^*$: f^* is disjunctive over V^*)
= {(9)}
   ($\forall V$: $r.V^*$: f is conjunctive over V)
= {under the assumption ($\forall V$:: $r.V \equiv r.V^*$) }
   ($\forall V$: $r.V$: f is conjunctive over V)
= {(14)}
   (f is r-conjunctive)   .

Hence, (16), and therefore (13) are achieved provided we introduce our junctivity types by restricting V's range to $r.V$ such that

(17)   ($\forall V$:: $r.V \equiv r.V^*$)   ,

i.e., r's such that $r.V$ is invariant under negation of all the predicates in bag V . (An example of an unacceptable r would be given by $r.V \equiv true \in V$.)

Because of the one-to-one correspondence between the elements of V and of V^* , any constraint on the cardinality of the bag V meets the requirement (17) on r .

A bag being "linear" means that its distinct elements can be arranged in a monotonic sequence; consequently the (finite or infinite) set of its distinct elements is denumerable. Linearity is invariant under the taking of the conjugate: there is a one-to-one correspondence between the distinct elements of V and those of V^* and the monotonicity of the sequence is maintained because negation is antimonotonic with respect to implication.

Remark For unclear reasons —and perhaps erroneously so— we did not pursue the generalization of linearity, viz., well-foundedness (see Chap. 9) with respect to implication or consequence: thanks to the antimonotonicity of the negation we have

$$(V \text{ is well-founded with respect to implication}) \equiv$$
$$(V^* \text{ is well-founded with respect to consequence}) \qquad .$$

(End of Remark.)

The ranges r —all satisfying (17)— we have chosen to highlight and the corresponding types of junctivity we have given a name are somewhat arbitrary. We distinguish the following types of junctivity:

● universally junctive, i.e., junctive over all V (of our junctivity types this is the only one that includes junctivity over the empty bag $\varnothing$)
● positively junctive, i.e., junctive over all non-empty V (so called because bags with cardinality zero are excluded)
● denumerably junctive, i.e., junctive over all non-empty V with denumerably many distinct predicates
● finitely junctive, i.e., junctive over all non-empty finite V (i.e., all non-empty V with a finite number of distinct predicates)
● $\cdots$ -continuous, i.e., junctive over all non-empty linear V (here we distinguish between "*and*-continuity" in the case of conjunctivity, and "*or*-continuity" in the case of disjunctivity; we have adopted the existing terminology)
● monotonic, i.e., junctive over all non-empty, finite, linear V (this name has to be justified by showing that this type of junctivity coincides with the notion of monotonicity as defined by (1)).

Let us discharge the last justification first, i.e. let us prove that for any f

(18)   (f is conjunctive over any non-empty, finite, linear V) $\equiv$
$$(\forall X,Y: [X \Rightarrow Y]: [f.X \Rightarrow f.Y])$$.

Proof The proof is by mutual implication.

LHS $\Rightarrow$ *RHS*   We observe for any f , conjunctive over all non-empty, finite, linear V , and any X , Y

$\quad [f.X \Rightarrow f.Y]$
$=$ {implication and conjunction}
$\quad [f.X \wedge f.Y \equiv f.X]$
$\Leftarrow$ {LHS; $\{X,Y\}$ is non-empty, finite and, if $[X \Rightarrow Y]$, linear}
$\quad [f.(X \wedge Y) \equiv f.X] \wedge [X \Rightarrow Y]$
$\Leftarrow$ {Leibniz}
$\quad [X \wedge Y \equiv X] \wedge [X \Rightarrow Y]$
$=$ {implication and conjunction; idempotence of $\wedge$ }
$\quad [X \Rightarrow Y]$.

$LHS \Leftarrow RHS$ The distinct predicates of a finite, linear bag can be ordered as a weakening sequence $Y.i$ for $0 \leqslant i < n$ with $n \geqslant 1$ if the bag is non-empty. Let this be done. We then observe

$\quad true$
$= \quad$ {above numbering convention}
$\quad (\forall i,j: \ 0 \leqslant i \leqslant j < n: \ [Y.i \Rightarrow Y.j])$
$\Rightarrow \quad$ {RHS with $X,Y := Y.i, \ Y.j$ and monotonicity of $\forall$ }
$\quad (\forall i,j: \ 0 \leqslant i \leqslant j < n: \ [f.(Y.i) \Rightarrow f.(Y.j)])$.

Hence, under the numbering convention and the assumption of RHS , the sequence $f.(Y.i)$ is weakening as well.

Finally, we observe

$\quad f.(\forall X: \ X \in V: \ X)$
$= \quad$ {see Remark below}
$\quad f.(\forall i: \ 0 \leqslant i < n: \ Y.i)$
$= \quad$ {(5, 114) $Y.i$ is weakening and $n \geqslant 1$ }
$\quad f.(Y.0)$
$= \quad$ {(5, 114) $f.(Y.i)$ is weakening and $n \geqslant 1$ }
$\quad (\forall X: \ X \in V: \ f.X)$.
$= \quad$ {see Remark below}
$\quad (\forall X: \ X \in V: \ f.X)$.

$\hfill$ (*End of Proof.*)

Remark In the above we appealed twice to a theorem that some readers will take for granted but that we prove for those that don't. For any bag of predicates V , any predicate transformer h , and any predicate-valued function Y we have —the range of i left unspecified and being understood—

(19) $(\forall X:: \ X \in V \ \equiv \ (\exists i:: \ [Y.i \ \equiv \ X])) \Rightarrow$
$\quad\quad [(\forall X: \ X \in V: \ h.X) \ \equiv \ (\forall i:: \ h.(Y.i))]$.

The antecedent of (19) is the formal rendering of the statement that the predicates $Y.i$ precisely span the predicates in V .

Proof We observe for any V , h , Y of the appropriate types

$\quad (\forall X: \ X \in V: \ h.X)$
$= \quad$ {trading}
$\quad (\forall X:: \ \neg(X \in V) \lor h.X)$
$= \quad$ {antecedent}
$\quad (\forall X:: \ \neg(\exists i:: \ [Y.i \equiv X]) \lor h.X)$
$= \quad$ {de Morgan}

$(\forall X:: \; (\forall i:: \; \neg[Y.i \equiv X]) \vee h.X)$

$= \quad \{ \; \vee \;$ distributes over $\; \forall \; \}$

$(\forall X:: \; (\forall i:: \; \neg[Y.i \equiv X]) \vee h.X))$

$= \quad \{$interchange of quantifications$\}$

$(\forall i:: \; (\forall X:: \; \neg[Y.i \equiv X] \vee h.X))$

$= \quad \{$(trading and) one-point rule$\}$

$(\forall i:: \; h.(Y.i))$.

(*End of Proof.*)

Negating in the consequent of (19) both sides and replacing h by $\neg h$ shows that

$$[(\exists X: \; X \in V: \; h.X) \equiv (\exists i:: \; h.(Y.i))]$$

would have been an equally acceptable consequent. Note that in the above proof we did not need to make any commitment about the type or the range of the dummy i . (*End of Remark.*)

The dual of (18) now follows straightforwardly:

$(f$ is disjunctive over any non-empty, finite, linear $V)$

$= \quad \{$(9) with $f := f*$ and $f** = f \; \}$

$(f*$ is conjunctive over any non-empty, finite, linear $V)$

$= \quad \{$(18) and (1) with $f := f* \; \}$

$(f*$ is monotonic)

$= \quad \{$(4)$\}$

$(f$ is monotonic)   .

Thus we have recognized monotonicity as the weakest type of junctivity. Moreover we have established that for this type of junctivity the distinction between conjunctivity and disjunctivity had disappeared; consequently, a predicate transformer that enjoys any type of junctivity is monotonic.

The following theorems —and their duals, whose formulation is left to the reader— are direct consequences of the definitions of our junctivity types. We have for any predicate transformer f

(20) $(f$ is universally conjunctive) $\Rightarrow$

$(f$ is positively conjunctive)

(21) $(f$ is positively conjunctive) $\Rightarrow$

$(f$ is denumerably conjunctive)

(22) $(f$ is denumerably conjunctive) $\Rightarrow$

$(f$ is finitely conjunctive) $\wedge (f$ is *and*-continuous)

(23) $(f$ is finitely conjunctive) $\vee (f$ is *and*-continuous) $\Rightarrow$

$(f$ is monotonic)   .

In short: in the order given, the types of conjunctivity form an almost weakening sequence of properties; only between finite conjunctivity and *and*-continuity no strength relation exists.

Theorem (22) can be strengthened, as is shown by the following

Theorem For any predicate transformer f

(24)   (f is denumerably conjunctive) $\equiv$
                    (f is finitely conjunctive) $\wedge$ (f is *and*-continuous)   .

Proof The proof is by mutual implication.

LHS $\Rightarrow$ *RHS* This is (22).

LHS $\Leftarrow$ *RHS* We observe for any predicate transformer f that is finitely conjunctive and *and*-continuous, and any sequence of predicates $X.i\,(0 \leqslant i)$:

$\quad f.(\forall i:\ 0 \leqslant i:\ X.i)$
$=\quad$ {pred. calc.: range of j is not empty}
$\quad f.(\forall i:\ 0 \leqslant i:\ (\forall j:\ i \leqslant j:\ X.i))$
$=\quad$ {unnesting}
$\quad f.(\forall i,j:\ 0 \leqslant i \wedge i \leqslant j:\ X.i)$
$=\quad$ {arithmetic, in particular transitivity of $\leqslant$ }
$\quad f.(\forall i,j:\ 0 \leqslant j \wedge 0 \leqslant i \leqslant j:\ X.i)$
$=\quad$ {nesting}
$\quad f.(\forall j:\ 0 \leqslant j:\ (\forall i:\ 0 \leqslant i \leqslant j:\ X.i))$
$=\quad$ { f is *and*-continuous and $(\forall i:\ 0 \leqslant i \leqslant j:\ X.i)\,(0 \leqslant j)$
            is a non-empty, strengthening sequence}
$\quad (\forall j:\ 0 \leqslant j:\ f.(\forall i:\ 0 \leqslant i \leqslant j:\ X.i))$
$=\quad$ { f is finitely conjunctive and $X.i\,(0 \leqslant i \leqslant j)$ is for $0 \leqslant j$
            a non-empty finite sequence}
$\quad (\forall j:\ 0 \leqslant j:\ (\forall i:\ 0 \leqslant i \leqslant j:\ f.(X.i)))$
$=\quad$ {as above, the other way round}
$\quad (\forall i:\ 0 \leqslant i:\ f.(X.i))$.

(*End of Proof.*)

The following theorem is included because we think its proof —which we owe to J.C.S.P. van der Woude— so nice.

Theorem For any predicate transformer f

(25)   (f is finitely conjunctive) $\wedge$ (f is *or*-continuous) $\Rightarrow$
            (f is *and*-continuous)   .

Proof   Under the truth of the antecedent of (25) we have to show for monotonic $X.i$ $(0 \leqslant i)$

$$[f.(\forall i:\ 0 \leqslant i:\ X.i) \equiv (\forall i:\ 0 \leqslant i:\ f.(X.i))]$$.

We distinguish two cases.

$X.i(0 \leqslant i)$ is weakening   The antecedent implies that f is monotonic. We observe for any monotonic f and weakening $X.i$ $(0 \leqslant i)$

$$f.(\forall i:\ 0 \leqslant i:\ X.i)$$
$$=\quad \{(5,\ 113),\ X.i\ (0 \leqslant i)\ \text{is weakening}\}$$
$$f.(X.0)$$
$$=\quad \{(5,\ 113),\ f.(X.i)\ (0 \leqslant i)\ \text{is weakening since}\ f\ \text{is monotonic}\}$$
$$(\forall i:\ 0 \leqslant i:\ f.(X.i))$$.

This was the not exciting part of the proof.

$X.i(0 \leqslant i)$ is strengthening   The antecedent of (25) implies that f is monotonic, and we are on account of (5, 108) therefore left with the proof obligation

(26) $$[f.(\forall i:\ 0 \leqslant i:\ X.i) \Leftarrow (\forall i:\ 0 \leqslant i:\ f.(X.i))]$$,

for strengthening $X.i$ $(0 \leqslant i)$ and an f that is finitely conjunctive and *or*-continuous.

Meeting the obligation of showing (26) is the exciting part of the proof. Reduced to its bare essentials, it consists of one definition and about a dozen steps. But in presenting just that irrefutable formal argument, we would pull several rabbits out of the magical hat. The proof is exciting because of the existence of heuristic considerations that quite effectively buffer these shocks of invention. For that reason, we shall develop this proof instead of just presenting it. To aid the reader in parsing the interleaved presentation of heuristic considerations and proof fragments, the latter will be indented. Here we go!

To begin with a general remark about the exploitation of *or*-continuity. The *or*-continuity of f states that

(27) $$[f.(\exists i::\ Y.i) \equiv (\exists i::\ f.(Y.i))]$$

for any monotonic sequence $Y.i$ $(0 \leqslant i)$. For a strengthening sequence $Y.i$ $(0 \leqslant i)$, just monotonicity of f suffices for (27) to hold, and for constant sequences $Y.i$ $(0 \leqslant i)$, (27) holds for any f . The relevant conclusion from these observations is that, if f's *or*-continuity is going to be exploited —and it is a safe assumption that it has to— a truly weakening sequence has to enter the picture.

Armed with this insight, we return to our demonstrandum (26). The simplest way of demonstrating an implication is to start at one side and then to repeatedly manipulate the expression (while either weakening or strengthening is allowed) until the other side is reached. So, let us try that. That decision being taken, at which side should we start?

Both sides are built from the "familiar" universal quantification and the "unfamiliar" application of f , about which our knowledge is limited, the only difference being that, at the two sides, they occur in opposite order. In such a situation, the side with the "unfamiliar" operation at the outside counts as the more complicated one, which is therefore the preferred starting point. In our case, it is the consequent

(28) $f.(\forall i:: X.i)$,

so let us start from there. The formal challenge of manipulating (28) while exploiting what we know about f should provide the heuristic guidance as to in which direction to proceed.

Rewriting (28) so as to exploit f's *or*-continuity would require to rewrite its argument $(\forall i:: X.i)$ as an existential quantification over a truly weakening sequence, but how to do that is not clear at all. So let us try to exploit at this stage f's finite conjunctivity, i.e., let us introduce a P and Q such that

(29) $[(\forall i:: X.i) \equiv P \wedge Q]$.

For one of the conjuncts, say P , we may choose any predicate implied by $(\forall i:: X.i)$; the law of instantiation tells us that any $X.j$ would do. (Note that this choice is less restrictive than it might seem: because $X.i$ $(0 \leqslant i)$ is strengthening, any finite conjunction of some $X.i$'s yields some $X.j$.) We could therefore consider for some j the introduction of a predicate Q constrained by

$[(\forall i:: X.i) \equiv X.j \wedge Q]$.

But the introduction of one predicate Q for one specific j is unlikely to do the job: for one thing, the universal quantifications in the demonstrandum don't change their value when the range $0 \leqslant i$ is replaced by $j < i$. This observation suggests, instead of the introduction of a single predicate Q a sequence, say $Y.j$ $(0 \leqslant j)$, constrained by

(30) $(\forall j:: [(\forall i:: X.i) \equiv X.j \wedge Y.j])$.

The introduction of the sequence $Y.j$ $(0 \leqslant j)$ will turn out to be the major invention in the proof under design. For the time being we don't define Y —as would be done immediately in a "bottom-up" proof— but only collect constraints on Y , of which (30) is the first one. We do so in the hope that, eventually, we can construct a Y that meets all the constraints.

A minor problem with the use of (30) as a rewrite rule is that it equates an expression not depending on j with one that formally does depend on j . The formal dependence on j that would thus be introduced can be eliminated by quantifying over it; because we are rewriting a consequent we use existential quantification because that yields a formally weaker expression than universal quantification (and, the weaker the consequent, the lighter the task ahead of us). In short, we propose to start our proof under design with

$$f.(\forall i::\ X.i)$$
$$=\quad \{(30)\text{ and range of } j \text{ non-empty}\}$$
$$(\exists j::\ f.(X.j \land Y.j))$$
$$=\quad \{\ f \text{ is finitely conjunctive}\}$$
(31) $(\exists j::\ f.(X.j) \land f.(Y.j))$.

So far, so good! We have not yet exploited f's *or*-continuity and we cannot do so before we have an existential quantification over a truly weakening sequence. We are not there yet, but we are getting close! In (31) we do have an existential quantification (be it, as yet, over a constant sequence) and, with $X.i \ (0 \leqslant i)$ a (truly) strengthening sequence, there is a fair chance that (30) permits a (truly) weakening sequence $Y.j \ (0 \leqslant j)$. So let us introduce the second constraint on Y

(32)   sequence $Y.j \ (0 \leqslant j)$ is weakening

as a next step towards the use of f's *or*-continuity, i.e., the use of (27) as a rewrite rule.

Comparison of the right-hand side of that rewrite rule (27) with (31) shows that we can apply the rewrite rule after we have succeeded in removing in (31) the first conjunct $f.(X.j)$ from the term. We cannot just omit it, as that would weaken the expression and, heading for an antecedent, we are not allowed to do that. We may strengthen it; in particular, strengthening it to something independent of j would allow us to take the constant conjunct outside the existential quantification of (31). In order to strengthen $f.(X.j)$ to something that is independent of j , we propose to quantify it universally over j . That is, at (31) we propose to continue our proof under design with

$$(\exists j::\ f.(X.j) \land f.(Y.j))$$
$$\Leftarrow\quad \{\text{instantiation, monotonicity of } \land\ ,\ \exists\ \}$$
$$(\exists j::\ (\forall i::\ f.(X.i)) \land f.(Y.j))$$
$$=\quad \{\ \land\ \text{ distributes over } \exists\ \}$$
$$(\forall i::\ f.(X.i)) \land (\exists j::\ f.(Y.j))$$
$$=\quad \{(27)\text{ and }(32), \text{ i.e., the use of } or\text{-continuity}\}$$
(33) $(\forall i::\ f.(X.i)) \land f.(\exists j::\ Y.j)$.

So far, so very good! Note, that the left conjunct of (33) is the antecedent of (26) that we are heading for! Again we cannot just omit the second conjunct, as that would weaken the expression; the second conjunct has to be subsumed —i.e., implied— by the first one. By the sight of it, we can equate (33) with its first conjunct on the strength of just the monotonicity of f and some implicative relation between X and Y —which will emerge as the third and last constraint on Y — . But be careful! If the range of i were empty, the first conjunct of (33) would yield *true*, whereas (33) would yield $f.(\exists j:: Y.j)$, and there is no reason to assume these equivalent. Somewhere along the completion of our formal argument, we have to exploit the non-emptiness of i's range! As we can do it immediately, let us do it immediately. In short, we propose to continue our proof under design at (33) with

$$(\forall i:: \ f.(X.i)) \wedge f.(\exists j:: \ Y.j)$$
$$= \quad \{\text{range of } i \text{ is non-empty}\}$$
$$(\forall i:: \ f.(X.i) \wedge f.(\exists j:: \ Y.j))$$
$$= \quad \{ \ f \text{ is monotonic and (34)}\}$$
$$(\forall i:: \ f.(X.i))$$

with, as our third and last constraint on Y ,

$$(34) \qquad (\forall i:: \ [X.i \Rightarrow (\exists j:: \ Y.j)]) \qquad .$$

But for the demonstration of the existence of Y , we have completed the proof in seven steps (six of which are equivalences). Now for the existence of Y .

In order to ease the satisfaction of (34) we define Y as the weakest solution of (30), i.e., we define for any j

$$(35) \qquad [Y.j \equiv (\forall i:: \ X.i) \vee \neg X.j] \qquad .$$

In order to verify that the first constraint, (30), is met, we observe for any j

$$X.j \wedge Y.j$$
$$= \quad \{(35)\}$$
$$X.j \wedge ((\forall i:: \ X.i) \vee \neg X.j)$$
$$= \quad \{ \ \wedge \text{ distributes over } \vee \ \}$$
$$(X.j \wedge (\forall i:: \ X.i)) \vee (X.j \wedge \neg X.j)$$
$$= \quad \{\text{instantiation and pred. calc.}\}$$
$$(\forall i:: \ X.i) \qquad .$$

In order to verify that $Y.j\ (0\leqslant j)$ is weakening we observe either that disjunction is monotonic and negation is antimonotonic, or in painstaking detail for any j and k

$$[Y.j\Rightarrow Y.k]$$
$$=\quad\{(35)\}$$
$$[(\forall i::\ X.i)\lor\neg X.j\Rightarrow(\forall i::\ X.i)\lor\neg X.k]$$
$$\Leftarrow\quad\{\text{pred. calc.}\}$$
$$[\neg X.j\Rightarrow\neg X.k]$$
$$=\quad\{\text{contra-positive}\}$$
$$[X.j\Leftarrow X.k]$$
$$\Leftarrow\quad\{\ X.i\ (0\leqslant i)\ \text{is strengthening}\}$$
$$j<k\qquad.$$

Finally, in order to verify that the last constraint on Y, (34), is met, we observe

$$(\exists j::\ Y.j)$$
$$=\quad\{(35)\}$$
$$(\exists j::\ (\forall i::\ X.i)\lor\neg X.j)$$
$$=\quad\{\ j\text{'s range is non-empty}\}$$
$$(\forall i::\ X.i)\lor(\exists j::\ \neg X.j)$$
$$=\quad\{\text{de Morgan}\}$$
$$(\forall i::\ X.i)\lor\neg(\forall i::\ X.i)$$
$$=\quad\{\text{Excluded Middle}\}$$
$$true\qquad.$$

And this concludes the exciting part.

(*End of Proof.*)

Junctivity theorems

The remaining task of this chapter is the development of a body of theorems to assist us in the establishment of the junctivity properties of given functions. For brevity's sake we shall concentrate on the conjunctivity properties of given functions: disjunctivity properties can be derived by studying the conjunctivity properties of the conjugate function.

It would be nice if our junctivity theory could be developed as simply as the punctuality theory. There, for a number of basic functions the punctuality in their argument(s) was postulated (for the equality) or proved (for the constant, the identity, the disjunction, and the negation), and then the Punctuality Theorem stated that, in all forms of composition, punctuality is preserved. But there are various reasons why, for our junctivity theory, such an easy way out has to remain a dream.

The first complication is that, while there was one notion of punctuality, there are different types of conjunctivity. The second complication is that —as we shall see in a moment— we can form from a universally conjunctive function a new function that is not conjunctive at all. The third complication is that we shall encounter a new type of problem, e.g., given the conjunctivity of g and h, how conjunctive is $g.X \wedge h.Y$, viewed as a function of the single argument (X,Y) ? We postpone for a while the discussion of this last type of problem.

The start is promising enough. We begin by establishing

(36) *Theorem* The identity function, i.e., the predicate transformer f given by

$$[f.X \equiv X] \quad \text{for all } X$$

is universally conjunctive.

Proof We observe for any V

$\quad f.(\forall X: X \in V: X)$
$= \quad \{\text{definition of } f \text{ with } X := (\forall X: X \in V: X) \}$
$\quad (\forall X: X \in V: X)$
$= \quad \{\text{definition of } f \}$
$\quad (\forall X: X \in V: f.X) \qquad .$

$\hspace{9cm}$ (*End of Proof.*)

So far, so good, but the trouble already starts with the constant function, for which we have

(37) *Theorem* A constant function, i.e., the predicate transformer f given for some Y by

(38) $\qquad [f.X \equiv Y] \quad \text{for all } X$

is positively conjunctive.

Proof We observe for any non-empty V

$\quad (\forall X: X \in V: f.X)$
$= \quad \{(38)\}$
$\quad (\forall X: X \in V: Y)$
$= \quad \{ V \text{ is non-empty}\}$
$\quad Y$
$= \quad \{(38) \text{ with } X := (\forall X: X \in V: X) \}$
$\quad f.(\forall X: X \in V: X) \qquad .$

$\hspace{9cm}$ (*End of Proof.*)

There is one constant function —viz., the constant *true* — that is universally conjunctive.

Theorem For a predicate transformer f given for some Y by (38)

(39) $(f$ is universally conjunctive$) \equiv [Y \equiv true]$.

Proof We observe for any f and Y satisfying (38)

   $(f$ is universally conjunctive$)$
= {definition of universal and positive conjunctivity}
   $(f$ is positively conjunctive$) \wedge (f$ is conjunctive over $\varnothing)$
= {(37)}
   $(f$ is conjunctive over $\varnothing)$
= {universal quantification with empty range yields *true* }
   $[f.true \equiv true]$
= {(38) with $X := true$ }
   $[Y \equiv true]$.

(End of Proof.)

Well, these were —to put it mildly— modest results. To give a little bit more direction to our investigations, we focus our attention on the weakest of all types of junctivity, viz., monotonicity. To restrict our space of investigation, we observe that we do not need to investigate those operators that, when applied to monotonic operands yield in general nonmonotonic results: whatever junctivity properties may be enjoyed by the components, they are not shared by the compositum. More precisely, if g and h are monotonic predicate transformers, $g.X \equiv h.X$, $g.X \Rightarrow h.X$, and $\neg g.X$ are in general *not* monotonic functions of X . Consequently, all junctivity properties are lost by the application of equivalence, implication, or negation.

Remark From a general point of view, this negative result seems at first sight rather disappointing. As will become clear in the next chapter, for the definition of programming language semantics our interest is confined to monotonic predicate transformers; consequently we won't encounter predicate transformers defined as equivalence, implication or negation. (*End of Remark.*)

Negative as the above conclusion may seem, it tells us where to look for more rewarding results: conjunction and disjunction, and their generalizations, universal and existential quantification. Let us deal with conjunction and universal quantification —including the "everywhere" operator— first; they seem to present the simpler situations.

(40) *Theorem* The "everywhere" operator is universally conjunctive, i.e., we have for any bag V of predicates

$$[[(\forall X:\ X \in V:\ X)] \equiv (\forall X:\ X \in V:\ [X])]$$.

Proof This is a restatement of (5, 88) about the interchange of universal quantifications; note that the range $X \in V$ is a boolean scalar.

(End of Proof.)

Since in the following theorem the set M may be finite or infinite and since conjunction may be identified with universal quantification over a finite range, the following theorem deals with conjunction and universal quantification at the same time.

(41) *Theorem* Let M be a set of predicate transformers, and let predicate transformer f in terms of M be given by

$$[f.X \ \equiv \ (\forall g:\ g \in M:\ g.X)] \quad \text{for all } X \ ;$$

then f enjoys each type of conjunctivity that is shared by all elements of M .

Proof This is proved by showing that f is conjunctive over any V over which all elements of M are conjunctive. For such a V we observe

$\quad f.(\forall X:\ X \in V:\ X)$
$=\quad$ {def. of f with $X := (\forall X:\ X \in V:\ X)$ }
$\quad (\forall g:\ g \in M:\ g.(\forall X:\ X \in V:\ X))$
$=\quad$ {all g conjunctive over V }
$\quad (\forall g:\ g \in M:\ (\forall X:\ X \in V:\ g.X))$
$=\quad$ {interchange of universal quantifications}
$\quad (\forall X:\ X \in V:\ (\forall g:\ g \in M:\ g.X))$
$=\quad$ {def. of f }
$\quad (\forall X:\ X \in V:\ f.X)$.

(End of Proof.)

This was a very general theorem, obtained at low cost. Let us now investigate the inheritance of conjunctivity in the case of disjunction. Here we may expect problems, so let us look at a simple case. Let f be given in terms of g and h by

$$[f.X \ \equiv \ g.X \lor h.X] \quad \text{for all } X \qquad .$$

Predicate transformer f being conjunctive over V then amounts to —the range $X \in V$ left understood—

$$[g.(\forall X::\ X) \lor h.(\forall X::\ X) \equiv (\forall X::\ g.X \lor h.X)] \qquad .$$

It is the right-hand side that is in general unmanageable: predicate calculus gives us in general no way of getting the disjunction out of the scope of a universal quantification. The previous chapter gives us two handles for special cases that might be manageable: (5, 78) and (5, 116). The first one is the distribution of $\vee$ over $\forall$, and suggests that we investigate, say, h being a constant function; the second one has a conjunction in the range which is as good as having a disjunction in the term. (See, however, the Confession below.)

(42) *Theorem* Let, for some predicate transformer g and some predicate Y , predicate transformer f be given by

$$[f.X \equiv g.X \vee Y] \quad \text{for all} \quad X \; ;$$

then f enjoys all conjunctivity properties enjoyed by g .

Proof Let g be conjunctive over V , and let the range $X \in V$ be understood. We then observe

$$
\begin{aligned}
&f.(\forall X :: X) \\
=\;& \{\text{def. of } f \text{ with } X := (\forall X :: X) \} \\
&g.(\forall X :: X) \vee Y \\
=\;& \{ \; g \text{ conjunctive over } V \} \\
&(\forall X :: g.X) \vee Y \\
=\;& \{ \; \vee \; \text{distributes over } \forall \} \\
&(\forall X :: g.X \vee Y) \\
=\;& \{\text{def. of } f \} \\
&(\forall X :: f.X) \qquad .
\end{aligned}
$$

(*End of Proof.*)

(43) *Theorem* Disjunction preserves *and*-continuity, i.e., let predicate transformer f be given in terms of the *and*-continuous predicate transformers g and h by

$$[f.X \equiv g.X \vee h.X] \quad \text{for all} \quad X \; ;$$

then f is *and*-continuous.

Proof Our proof obligation is to show for *and*-continuous g and h , and any monotonic sequence $X.i \, (0 \leqslant i)$ and f given as above

$$[f.(\forall i: \; 0 \leqslant i: \; X.i) \equiv (\forall i: \; 0 \leqslant i: \; f.(X.i))] \qquad .$$

Let it be understood for dummies i and j that they range over the natural numbers. To begin with, we observe

$(g.(X.i) \lor h.(X.j)$ is weakening in i and j or is strengthening in i and $j)$

$\Leftarrow \quad \{ \lor$ is monotonic in both arguments$\}$

$(g.(X.i)$ and $h.(X.i)$ are both weakening in i or both strengthening in $i)$

$\Leftarrow \quad \{ g$ and h are *and*-continuous, hence monotonic$\}$

$(X.i$ is a monotonic sequence)   .

After these preliminaries we observe

$f.(\forall i:: \ X.i)$

$= \quad \{$def. of f with $X := (\forall i:: \ X.i)$ and change of dummy$\}$

$g.(\forall i:: \ X.i) \lor h.(\forall j:: \ X.j)$

$= \quad \{ g$ and h *and*-continuous; $X.i$ monotonic$\}$

$(\forall i:: \ g.(X.i)) \lor (\forall j:: \ h.(X.j))$

$= \quad \{ \lor$ distributes over $\forall$ $\}$

$(\forall i:: \ g.(X.i) \lor (\forall j:: \ h.(X.j)))$

$= \quad \{ \lor$ distributes over $\forall$ $\}$

$(\forall i:: \ (\forall j:: \ g.(X.i) \lor h.(X.j)))$

$= \quad \{(5, 116)$ and above preliminary$\}$

$(\forall i:: \ g.(X.i) \lor h.(X.i))$

$= \quad \{$def. of f with $X := X.i$ $\}$

$(\forall i:: \ f.(X.i))$.

(*End of Proof.*)

Confession As revealed by the structure of the proof, the reason we gave for looking at (5, 116) is most unconvincing. The reader may safely assume that the preceding chapter was written with the current one in mind. We discovered relatively late in the development that we could be much more explicit about heuristics than we had been able before. In this case, we allowed ourselves to be carried away by our enthusiasm. (*End of Confession.*)

By repeated application of the last theorem we conclude that *and*-continuity is preserved by finite disjunctions or, what amounts to the same thing, existential quantification over a finite range. Similarly, *or*-continuity is preserved under universal quantification over a finite range. The wide range of circumstances under which continuity is preserved —in a moment we shall encounter yet another one— is one of the reasons why the concept of continuity has attracted attention; the complementary reason is of course that —as we shall see later— the concept can be exploited. In the next chapter, where we start dealing with programming language semantics, we shall encounter the choice whether to admit infinite guarded command sets, and, as a universal quantification over the members of that set enters the game, we shall see that that choice boils down to whether to forsake *or*-continuity. We are beginning to see the mathematical reasons why such a decision of language design should not be taken lightly.

The restriction of our interest to at least monotonic functions ruled out function formation by equivalence, implication, and negation. That could be viewed as a negative conclusion. In a more positive mood we can explore what we can do thanks to the fact that all functions of interest are monotonic. This is done in the following two theorems; the first is primarily a stepping-stone for the second.

(44) *Theorem* Let V be a bag of predicates and let h be a monotonic predicate transformer. Let W be the bag obtained by replacing each predicate X in V by $h.X$. Then W is of V's "junctivity type", i.e., the restrictions concerning cardinality and linearity that are met by V are met by W as well.

Proof If V is non-empty/denumerable/finite, then W is by its construction non-empty/denumerable/finite, and these three are the only cardinality constraints that play a rôle in the definition of the junctivity types. If the (distinct) elements of V can be written as a monotonic sequence $X.i\,(0 \leqslant i)$, then, by construction, $h.(X.i)\,(0 \leqslant i)$ contains all distinct elements of W , and moreover, because also h is monotonic, the sequence $h.(X.i)\,(0 \leqslant i)$ is monotonic. Hence, if V is linear, so is W .

(End of Proof.)

And now we are ready to explore another way of forming functions from functions, viz., functional composition. This leads to the beautiful

(45) *Theorem* Functional composition is junctivity preserving, i.e., let, for some predicate transformers g and h , predicate transformer f be given by

$$[f.X \equiv g.(h.X)] \quad \text{for all } X \ ;$$

then f enjoys each junctivity property shared by g and h .

Proof We can confine ourselves to monotonic g and h , because otherwise there are no shared junctivity properties, in which case the theorem vacuously holds. We give the proof for any conjunctivity property.

Let g and h be conjunctive over predicate bags of some junctivity type and let V be of that type. Let W be given by —see Note—

$$W = \{X\colon\ X \in V\colon\ h.X\} \qquad ;$$

because h is monotonic, W is on account of (44) of V's junctivity type.

And now we observe

$f.(\forall X: X \in V: X)$
$=$ {def. of f with $X := (\forall X: X \in V: X)$ }
$g.(h.(\forall X: X \in V: X))$
$=$ { h is conjunctive over V }
$g.(\forall X: X \in V: h.X)$
$=$ {relation between V and W , (46) with $p, q := h$, identity}
$g.(\forall Y: Y \in W: Y)$
$=$ { g is conjunctive over W }
$(\forall Y: Y \in W: g.Y)$
$=$ {relation between V and W , (46) with $p, q := h, g$ }
$(\forall X: X \in V: g.(h.X))$
$=$ {def. of f }
$(\forall X: X \in V: f.X)$.

(*End of Proof.*)

Note The set notation using the braces is formally given by: for all p, V

$$(\forall Y:: Y \in \{X: X \in V: p.X\} \equiv$$
$$(\exists X: X \in V: [p.X = Y]))$$.

For the sake of completeness, we observe for any Y , V

$Y \in \{X: X \in V: X\}$
$=$ {above definition with $p :=$ identity function}
$(\exists X: X \in V: [X = Y])$
$=$ {trading and one-point rule}
$Y \in V$,

so that, indeed, we have $\{X: X \in V: X\} = V$.

Furthermore we have for all V , W , p , q

(46) $W = \{X: X \in V: p.X\} \Rightarrow$
$[(\forall Y: Y \in W: q.Y) \equiv (\forall X: X \in V: q.(p.X))]$.

Proof We observe for any V , W , p , q

$(\forall Y: Y \in W: q.Y)$
$=$ {antecedent of (46)}
$(\forall Y: Y \in \{X: X \in V: p.X\}: q.Y)$
$=$ {definition of braces for set notation}
$(\forall Y: (\exists X: X \in V: [p.X = Y]): q.Y)$
$=$ {trading}

$(\forall Y: (\exists X: X \in V \wedge [p.X = Y]: \text{true}): q.Y)$

$=$ {trading and de Morgan}

  $(\forall Y:: (\forall X: X \in V \wedge [p.X = Y]: \text{false}) \vee q.Y)$

$=$ { $\vee$ distributes over $\forall$; pred. calc.; unnesting}

  $(\forall X,Y: X \in V \wedge [p.X = Y]: q.Y)$

$=$ {nesting and one-point rule}

  $(\forall X: X \in V: q.(p.X))$

(*End of Proof.*)

(*End of Note.*)

*   *   *

We now must draw the reader's attention to a dilemma that we face when dealing with what is loosely called "a function of more arguments". For simplicity's sake, we shall illustrate the dilemma with, and develop our theorems for, "functions of two arguments", leaving the generalization from pairs to n-tuples to the reader.

We take as starting point what is usually written as "$f(x,y)$" . Adopting the convention of explicity indicating functional application by a full stop ($=$ period), we would write "$f.(x,y)$" , which admits of only one interpretation: f applied to the argument (x,y) —which happens to be of type "pair"— .

The alternative —pioneered by the logician H. B. Curry— is to consider a (higher order) function F that, when applied to x , yields a new function that, in turn, can be applied to y to yield $f.(x,y)$. With functional application left-associative —i.e., "$F.x.y$" standing for "$(F.x).y$"— the relation between f and F is

  $[f.(x,y) = F.x.y]$ for all x , y .

Standard example Going from the infix " $+$ " to the prefix "*add*" by defining

  $add.x.y = x + y$,

we get a higher-order function with *add*.1 and *add*.(-1) as very familiar values: they are the successor function and predecessor function, respectively. (*End of Standard example.*)

The study of the relationship between the above f and F and the development of the notational equipment needed to express the one in terms of the other gave rise to the theory known as "Combinatorial Logic", a nice theory that we need not be concerned with here. We only face the dilemma which of the two formats to adopt, as each has its advantages and disadvantages.

Remark   The informal mathematician, as said, writes $f(x,y)$ and, his functional application being invisible, he hardly notices the difference. But in a formal environment such as a programming language, one has to be explicit: it must be absolutely clear whether one introduces a name like f or like F . The problem is acute for the users of functional programming languages. Consequently, such programming languages are in constant danger of getting burdened with special features to express in terms of functions of the one format what would have been easily expressible in terms of the corresponding function of the other format. (*End of Remark.*)

For our current investigations, the format of the above f is the most convenient, i.e., we shall consider functions of predicate pairs. The format of the above F would have had two disadvantages. Firstly, we would be faced with a formula in which the two predicates occur very asymmetrically, secondly, we would be forced to introduce the notion of junctivity for higher-order functions. We could meet those challenges, but it seems simpler to avoid them.

The price we have to pay for the introduction of predicate pairs is our willingness to get accustomed to manipulating them. We should pay that price gladly: firstly, the price is low because the rules of manipulation are very simple, and, secondly, the significance of the notion of a predicate pair is by no means restricted to junctivity considerations.

In denoting a predicate pair, we shall follow the mathematical custom and write "(X,Y)" —rather than "*pair.X.Y*"— for any predicates X and Y. The comma separating the two components X and Y has a very low binding power, lower than all boolean operators. Note that the surrounding parentheses are not optional.

The first rule about pair-forming is

(47) $[(X,Y)] \equiv [X] \wedge [Y]$.

The others state that the boolean operators distribute over pair-forming:

(48) $[(X , Y) \equiv (X' , Y') \equiv (X \equiv X' , Y \equiv Y')]$
            $[(X , Y) \vee (X' , Y') \equiv (X \vee X' , Y \vee Y')]$
            $[(X , Y) \wedge (X' , Y') \equiv (X \wedge X' , Y \wedge Y')]$
            $[(X , Y) \Rightarrow (X' , Y') \equiv (X \Rightarrow X' , Y \Rightarrow Y')]$
            $[\neg(X , Y) \equiv (\neg X , \neg Y)]$
            $[(\forall i:: (X.i , Y.i)) \equiv ((\forall i:: X.i) , (\forall i:: Y.i))]$
            $[(\exists i:: (X.i , Y.i)) \equiv ((\exists i:: X.i) , (\exists i:: Y.i))]$.

All this is —see Chap. 1— in accordance with viewing a predicate pair as a predicate on a doubled space, with viewing the "everywhere" operator as universal quantification over the underlying space, and with the point-wise application of the operators and the quantifications.

Now we turn our attention to functions, whose values or whose arguments are predicate pairs. First we deal with the former.

(49) *Theorem*   Pair-forming is junctivity preserving, i.e., predicate transformer f , given for some g and h by

$$[f.X \equiv (g.X \, , \, h.X)] \quad \text{for all} \ \ X$$

enjoys all junctivity properties shared by g and h .

Proof We shall prove the theorem for any conjunctivity property shared by g and h . For the disjunctivity properties the theorem then follows from (13) and the fact that

$$[f^*.X \equiv (g^*.X \, , \, h^*.X)] \qquad .$$

To show f's inheritance of conjunctivity, we observe for any V over which g and h are conjunctive —the range $X \in V$ being understood—

$\quad f.(\forall X:: \ X)$
$= \quad \{\text{def. of} \ f \ \text{with} \ X := (\forall X:: \ X) \ \}$
$\quad (g.(\forall X:: \ X) \, , \, h.(\forall X:: \ X))$
$= \quad \{ \ g \ \text{and} \ h \ \text{conjunctive over} \ V \ \}$
$\quad ((\forall X:: \ g.X) \, , \, (\forall X:: \ h.X))$
$= \quad \{ \ \forall \ \text{distributes over} \ (\, , \,) \ \}$
$\quad (\forall X:: \ (g.X \, , \, h.X))$
$= \quad \{\text{def. of} \ f \ \}$
$\quad (\forall X:: \ f.X) \qquad .$

(End of Proof.)

A sort of inverse of the previous theorem is

(50) *Theorem* The selector functions are universally junctive, i.e., predicate transformers *left* and *right* , given by

(51) $[left.(X,Y) \equiv X]$ for all (X,Y)

(52) $[right.(X,Y) \equiv Y]$ for all (X,Y)

are universally junctive.

Proof For reasons of symmetry it suffices to prove the theorem for *left* . Because *left* and *left** are the same function, we need consider only universal conjunctivity.

We observe, for W any bag of predicate pairs, and the range $(X,Y) \in W$ being understood,

$left.(\forall X,Y:: (X,Y))$
= $\quad\{ \forall$ distributes over $(,) \}$
$left.((\forall X,Y:: X) , (\forall X,Y:: Y))$
= $\quad\{$def. of $left$ with $X,Y := (\forall X,Y:: X) , (\forall X,Y:: Y) \}$
$(\forall X,Y:: X)$
= $\quad\{$def. of $left \}$
$(\forall X,Y:: left.(X,Y))$.

$\hspace{8cm}$ *(End of Proof.)*

Functions *left* and *right* enable us to write any expression in X and Y as an expression in (X,Y) . We shall show how this is exploited in the proof of

(53) *Theorem* Let, for some predicate transformers g and h , predicate transformer f be given by

$$[f.(X,Y) \equiv g.X \wedge h.Y] \quad \text{for all } X , Y \hspace{2cm} ;$$

then f enjoys all conjunctivity properties shared by g and h .

Proof To begin with we observe

$f.(X,Y)$
= $\quad\{$def. of $f \}$
$g.X \wedge h.Y$
= $\quad\{$def. of $left$ and $right \}$
$g.(left.(X,Y)) \wedge h.(right.(X,Y))$
= $\quad\{$def. of functional composition$\}$
$(g \circ left).(X,Y) \wedge (h \circ right).(X,Y)$.

But now f has been rewritten as a conjunction of two functions of its argument, and according to theorem (41), f enjoys each type of conjunctivity shared by $g \circ left$ and $h \circ right$. Because —theorem (45)— functional composition is junctivity preserving, f enjoys therefore each type of conjunctivity shared by g , h , *left* , and *right* . Because the latter two —theorem (50)— are universally junctive, the conclusion follows.

$\hspace{8cm}$ *(End of Proof.)*

In the same vein we can deduce from theorem (43)

(54) *Theorem* Let for some *and*-continuous predicate transformers g and h , predicate transformer f be given by

$$[f.(X,Y) \equiv g.X \vee h.Y] \quad \text{for all } X , Y ;$$

then f is *and*-continuous.

The proof is left to the reader.

Up till now we have looked at the junctivity properties of functions of one well-identified argument. Either that one well-identified argument was a single predicate —as in "$f.X$" — and we would compare quantifications over a bag V of single predicates— or that one well-identified argument was a predicate pair —as in "$f.(X,Y)$" — and we would compare quantifications over a bag W of predicate pairs.

Now it is time to face the fact that $f.(X,Y)$ admits of two other functional views besides a function of its total argument:

● we can view it as "a function in the first component", i.e., for any fixed Y we can view $f.(X,Y)$ as a function in the single predicate X ;
● we can view it as "a function in the second component", i.e., for any fixed X we can view $f.(X,Y)$ as a function in the single predicate Y .

The question is now how the junctivity of the original f is related to the junctivity of the latter two functions of a single predicate, or —to use the jargon— how "f's junctivity in its total argument" is related to "f's junctivity in the first and second component".

There are two ways of expressing formally what junctivity in a component means. For instance, "f is conjunctive over V in the first component" means that we have for any Y .

● $[f.((\forall X: X \in V: X),Y) \equiv (\forall X: X \in V: f.(X,Y))]$, or

● "f' is conjunctive over V" with f' defined by

(55) $[f'.X \equiv f.(X,Y)]$ for any X .

To prepare the investigation to what extent junctivity in the total argument implies junctivity in the components, we first establish the

Theorem Let, for some predicate Y , predicate transformer k be given by

(56) $[k.X \equiv (X,Y)]$ for all X ;

then k is positively junctive.

Proof With g and h defined by

$$[g.X \equiv X] \quad \text{and} \quad [h.X \equiv Y] \quad \text{for any } X$$

we observe

$$
\begin{aligned}
& k.X \\
= \quad & \{\text{def. of } k \} \\
& (X, Y) \\
= \quad & \{\text{def. of } g \text{ and } h \} \\
& (g.X , h.X) \qquad .
\end{aligned}
$$

Hence —theorem (49)— , k enjoys all junctivity properties shared by g and h . Because, furthermore, g (the identity function) —theorem (36)— is universally junctive and h (a constant function) —theorem (37)— is positively junctive, k is positively junctive.

(End of Proof.)

And now we are ready for the following general result.

(57) *Theorem* With the exception of universal junctivity, a function is as junctive in its components as it is in its total argument.

Proof We have to show that, with f' expressed in terms of f as in (55), f' is as junctive as f , with the exception of universal junctivity. To this end we observe that for some Y

$$
\begin{aligned}
& f'.X \\
= \quad & \{(55)\} \\
& f.(X, Y) \\
= \quad & \{\text{with } k \text{ as defined by (56)}\} \\
& f.(k.X) \qquad .
\end{aligned}
$$

Because —theorem (45)— functional composition is junctivity preserving and —the previous theorem— k enjoys all junctivities other than universal junctivity, the theorem now follows.

(End of Proof.)

And now we face the inverse question: let for a function of a predicate pair be given how junctive it is in its components, how junctive in its total argument can we conclude it to be? In this direction, the inheritance is much weaker.

(58) *Theorem* A function of a predicate pair that is monotonic in both components, *and*-continuous in both components, or *or*-continuous in both components, is so in its total argument.

Proof   We give the proof for monotonicity and *and*-continuity. Let
$(X.i , Y.i) (0 \leqslant i)$ be a finite or infinite, monotonic sequence. Then, the
sequences $X.i$ and $Y.i$ —ranges from now on to be understood— are both
weakening or both strengthening and, if finite, of the same length. Let f be
conjunctive over $X.i$ in the first component and conjunctive over $Y.i$ in the
second component. Then we observe

$f.(\forall i:: (X.i , Y.i))$

= { $\forall$ distributes over pair-forming; renaming a dummy}

$f.((\forall i:: X.i) , (\forall j:: Y.j))$

= { f conjunctive over $X.i$ in 1st component}

$(\forall i:: f.(X.i , (\forall j:: Y.j)))$

= { f conjunctive over $Y.j$ in 2nd component}

$(\forall i:: (\forall j:: f.(X.i , Y.j)))$

= {unnesting}

$(\forall i,j:: f.(X.i , Y.j))$

= {(5, 116); because f is monotonic in both its components and $X.i$ and
$Y.j$ are both weakening or both strengthening, $f.(X.i , Y.j)$ is
weakening in both i and j or strengthening in both i and j }

$(\forall i:: f.(X.i , Y.i))$.

(End of Proof.)

Remark   Note that our earlier theorem (54), whose proof we left to the reader,
can also be proved by an appeal to the above inheritance theorem.

(End of Remark.)

For the ease of retrieval, we give the following corollary of our last two
theorems.

Corollary   For a function f of an n-tuple of predicates we have

(59)   (f is monotonic in its total argument) $\equiv$
(f is monotonic in all the components)

(60)   (f is *and*-continuous in its total argument) $\equiv$
(f is *and*-continuous in all the components)

(61)   (f is *or*-continuous in its total argument) $\equiv$
(f is *or*-continuous in all the components)   .

The first one is rather obvious and easily demonstrated in isolation. The last
two are less obvious, but give a very clear hint why continuity —a rather
complicated property to define!— is a significant notion. An important
consequence of the corollary is that we can now talk about monotonic

functions, *and*-continuous functions, and *or*-continuous functions without distinguishing between junctivity in the total argument and junctivity in all the components.

Some justifying examples

We have introduced a whole hierarchy of junctivity types; this raises the question of how meaningful all these distinctions are. Do there really exist, for each junctivity property, functions enjoying it but not enjoying the next stronger one? This can be settled by examples.

We have also given quite a number of theorems; but a theorem raises in general the question of whether we should have replaced it by some stronger one. This can be settled by an example that refutes such a proposed stronger version.

This section is devoted to some of such examples, showing that our distinctions make some sense or that our theorems are not unnecessarily weak. The set of examples given below does not have the slightest claim of being complete (in whatever sense).

● Do there exist functions at all that satisfy our strongest constraints, i.e., being universally conjunctive or universally disjunctive? (If not, universal junctivity would, as a concept, not make much sense.) The answer is "Yes". The identity function —i.e., f given by $[f.X \equiv X]$ — settles the question.

● Is the distinction between universal and positive junctivity meaningful? The answer is "Yes", as follows from the existence of a function that is positively conjunctive, but not universally so, e.g., f given by $[f.X \equiv false]$.

● Is the distinction between positive and denumerable junctivity meaningful? The answer is "Yes", as follows from the existence of a function that is denumerably disjunctive but not positively so. Consider a space of a nondenumerable set of points and define the predicate transformer f by

$$[f.X \equiv (X \text{ holds in a nondenumerable set of points})]$$.

(On a space of a denumerable set of points, f would not be very interesting: it would be the constant function *false* .) We now show that f is denumerably disjunctive, but not positively so. Denumerable disjunctivity amounts to the fact that for any denumerable set V of predicates

(62) $[f.(\exists X:\ X \in V:\ X) \equiv (\exists X:\ X \in V:\ f.X)]$

or, equivalently

$((\exists X:\ X \in V:\ X)$ holds in a nondenumerable set of points) $\equiv$
$(\exists X:\ X \in V:\ (X$ holds in a nondenumerable set of points))   ,

which, for denumerable V , is a theorem of set theory. (Viz., the union of a denumerable set of sets is nondenumerable if and only if at least one of those sets is nondenumerable.) To show that f is not positively disjunctive, we take a non-empty V for which (62) is *false* . It suffices to take for V the set of all "point predicates", i.e., all predicates that hold in a single point of space. For that V we observe

$$f.(\exists X:\ X \in V:\ X)$$
$$=\quad \{\text{because } V \text{ contains all point predicates}\}$$
$$f.true$$
$$=\quad \{\text{def. of } f \text{ and nondenumerability of space}\}$$
$$true$$
$$\neq\quad \{\text{predicate calculus}\}$$
$$(\exists X:\ X \in V:\ false)$$
$$=\quad \{\text{def. of } f\ ,\ V\ , \text{ and the singleton set is denumerable}\}$$
$$(\exists X:\ X \in V:\ f.X)\qquad .$$

• Is the distinction between denumerable junctivity, finite junctivity, and continuity meaningful? The answer is "Yes". Because of theorem (24)

$$(f \text{ is denumerably conjunctive}) \equiv$$
$$(f \text{ is finitely conjunctive}) \wedge (f \text{ is } and\text{-continuous})\qquad ,$$

it suffices to show (i) a finitely conjunctive f that is not *and*-continuous, and (ii) an *and*-continuous f that is not finitely conjunctive.

(i) Let $P.i\ (0 \leqslant i)$ be a strengthening sequence, such that   —ranges being understood to be over the naturals—

(*) $\neg(\exists i::\ [P.i \equiv (\forall j::\ P.j)])$.

(This means that $P.i$ is an "ever strengthening sequence"; such sequences are quite thinkable, e.g., $[P.i \equiv i \leqslant n]$.) Consider now predicate transformer f given by

$$[f.X \equiv (\exists i::\ [P.i \Rightarrow X])]\quad \text{for all } X\qquad .$$

Now observe for any X , Y

$$f.(X \wedge Y)$$
$$=\quad \{\text{def. of } f \text{ with } X := X \wedge Y\ \}$$
$$(\exists i::\ [P.i \Rightarrow X \wedge Y])$$
$$=\quad \{\text{pred. calc.}\}$$
$$(\exists i::\ [P.i \Rightarrow X] \wedge [P.i \Rightarrow Y])$$
$$=\quad \{\text{dual of (5, 116)}; [P.i \Rightarrow X] \wedge [P.j \Rightarrow Y] \text{ is weakening in both } i \text{ and } j\ \}$$
$$(\exists i,j::\ [P.i \Rightarrow X] \wedge [P.j \Rightarrow Y])$$
$$=\quad \{\text{nesting; } \wedge \text{ distributes over } \exists\ , \text{ twice}\}$$
$$(\exists i::\ [P.i \Rightarrow X]) \wedge (\exists j::\ [P.j \Rightarrow Y])$$
$$=\quad \{\text{def. of } f\ , \text{ as is and with } X := Y\ \}$$
$$f.X \wedge f.Y\qquad ;$$

hence f is finitely conjunctive. Next we observe

$$f.(\forall j:: \ P.j)$$
$$= \quad \{\text{def. of } f \text{ with } X := (\forall j:: \ P.j) \ \}$$
$$(\exists i:: \ [P.i \Rightarrow (\forall j:: \ P.j)])$$
$$= \quad \{\text{since } -\text{pred. calc.} - \ [P.i \Leftarrow (\forall j:: \ P.j)] \ \}$$
$$(\exists i:: \ [P.i \equiv (\forall j:: \ P.j)])$$
$$= \quad \{(*)\}$$
$$false \qquad .$$

On the other hand, we observe

$$(\forall j:: \ f.(P.j))$$
$$= \quad \{\text{def. of } f \text{ with } X := P.j \ \}$$
$$(\forall j:: \ (\exists i:: \ [P.i \Rightarrow P.j]))$$
$$= \quad \{\text{since } [P.j \Rightarrow P.j] \ \}$$
$$true \qquad .$$

Combining the two observations, we conclude

$$[f.(\forall j:: \ P.j) \not\equiv (\forall j:: \ f.(P.j))] \qquad .$$

Hence, $P.j$ being a monotonic sequence, f is not *and*-continuous. And that concludes example (i).

(ii) Now we have to construct an *and*-continuous f that is not finitely conjunctive. We can do so with the aid of substitution, which is a predicate transformer that will be discussed in the next and last section of this chapter. It can be defined in the special case of predicates on a state space. The predicates then correspond to boolean expressions in the variables spanning the state space, and the "everywhere" operator corresponds to universal quantification over all those variables.

Let n be one of the variables spanning the state space and let X be a predicate on the state space, i.e., a boolean expression with possibly n as one of its global ($=$ free) variables. Then "(X with $n := E$)" will be used here to denote the result obtained by substituting E for n in X . We shall see in the next section that substitution is a universally junctive predicate transformer.

And now we are ready to construct the predicate transformer we were looking for: let f be defined by

(63) $[f.X \equiv (X \text{ with } n := 0) \vee (X \text{ with } n := 1)]$ for all X .

Substitution, being universally junctive, is *and*-continuous; theorem (43) tells us that the disjunction preserves *and*-continuity. Hence, the above defines an *and*-continuous f .

For the predicates $n = 0$ and $n = 1$ we observe

$f.(n = 0 \ \wedge \ n = 1)$
$=$ $\{ \ 0 \neq 1 \ \}$
$f.false$
$=$ $\{$def. of f with $X := false \ \}$
$(false$ with $n := 0) \vee (false$ with $n := 1)$
$=$ $\{$substitution$\}$
$false \vee false$
$=$ $\{ \ \vee$ is idempotent$\}$
$false$.

However,

$f.(n = 0) \wedge f.(n = 1)$
$=$ $\{$def. of f with $X := n = 0$ and $X := n = 1 \ \}$
$(((n = 0)$ with $n := 0) \vee ((n = 0)$ with $n := 1)) \wedge$
$(((n = 1)$ with $n := 0) \vee ((n = 1)$ with $n := 1))$
$=$ $\{ \ 0 = 0, 1 = 1, 0 \neq 1 \ \}$
$(true \vee false) \wedge (false \vee true)$
$=$ $\{$pred. calc.$\}$
$true$.

Combining these observations, we conclude that our *and*-continuous f is not finitely conjunctive. And that concludes example (ii).

● Is the distinction between finite junctivity or continuity versus monotonicity meaningful? The answer is "Yes". We shall construct a monotonic predicate transformer f that is neither finitely conjunctive nor *and*-continuous. It is really a generalization of the previous example (ii): we generalize it so as to destroy *and*-continuity as well. Consider the predicate transformer f given by

$$[f.X \equiv (\exists i :: (X \text{ with } n := i))] \quad \text{for all } X \quad .$$

We leave it to the reader to show that this f is monotonic, but not finitely conjunctive and not *and*-continuous. (Hint: to refute *and*-continuity, consider the strengthening sequence $P.j$, given by $[P.j \equiv n \geq j]$ for all j .)

● Is the notion of monotonicity meaningful? The answer is "Yes", for there are also predicate transformers that are not monotonic, e.g., f given by $[f.X \equiv \neg X]$.

So much for the meaningfulness of our junctivity distinctions.

*   *   *

Let us now have a look at some of our theorems and convince ourselves that they are not unnecessarily weak.

Theorem (41) guarantees "each type of conjunctivity that is shared by all elements of M ". That it is in general impossible to guarantee more conjunctivity is shown by the case that M has only one element g : in that case f equals g . The question how much disjunctivity is inherited from the conjuncts is the dual of the question how much conjunctivity is inherited from disjuncts, but the answer is "In general not much.": the disjuncts in (63) are universally conjunctive, but the f there defined is not even finitely so, and thereby justifies the constraints of theorem (43).

Theorem (45) states that functional composition is junctivity preserving; could the composition be of a stronger junctivity than one of the functions composed? In general hardly so, as is shown by taking for one of the two the —universally junctive!— identity function; the compositum then equals the other function.

In theorem (57), universal junctivity is excluded from the inheritance, and rightly so, as shown by the following example:

$$[f.(X,Y) \equiv X \vee Y] \qquad \text{and} \qquad [f'.X \equiv f.(X,Y)]$$

Here f is universally disjunctive, but f' in general only positively so.

The same f can be used to show that the restrictions in theorem (58) are not void: f is universally conjunctive in both components —theorem (42) with $g :=$ the identity function— but, as the reader can verify, not even finitely conjunctive in its total argument.

And this concludes our justifying examples.

Substitution as predicate transformer

A main activity in this chapter so far has been to compose new predicate transformers from given ones and to derive properties of the compositum from properties of the given components. But from where did we get those given components? In this section we shall present a method of constructing such component predicate transformers.

The method is in two ways very powerful. Firstly, it enables us to generate infinitely many predicate transformers of a dazzling variety; consequently, predicates and their transformers form a universe of discourse that can be very rich indeed. Secondly, the predicate transformers that are constructed by this method are all universally junctive, and that puts the preceding part of this chapter in proper perspective: the ways of composing predicate transformers so as to form new ones is the major source of loss of junctivity properties.

In our application to programming language semantics, this method of constructing predicate transformers is, in essence, even our only way of constructing our basic building blocks. The method being so powerful and seemingly basic, the reader may wonder why we did not introduce it right at the beginning of this chapter instead of postponing it until the very last section of this chapter. The reason is very simple: the method only works for predicates more specific than the general theory requires. The method requires the predicates to be predicates on a state space, i.e., the predicates can be identified with boolean expressions in the variables —or, if you prefer: Cartesian coordinates— spanning the state space and the "everywhere" operator can be identified with universal quantification over all those variables. (The reader may verify that this specialization is compatible with all the postulates of our calculus of boolean structures.)

Let m and n be the two variables of the state space; bearing that in mind, $[X]$ may be read as $(\forall m, n:: X)$, in which it is to be understood that, in general, X depends on both m and n .

Remark We can restrict ourselves without loss of generality to a state space spanned by two variables: for any dichotomy of a larger number of state variables, we can always lump the variables of the one set together into a variable m and those of the other set into a variable n . There is no need to denote them by $\overline{m}$ and $\bar{n}$, just to remind us that each of them "really" stands for a tuple of variables; on the contrary, the sooner this "reality" is forgotten, the better, for it is totally irrelevant. Compared with this conceptual simplification, the typographical simplification of omitting the overlining is only a minor one. (*End of Remark.*)

Let E be an expression of the same type as n , where, in general, E also depends on m and n . For given n , E , and X , the predicate of interest is the one obtained by replacing all global (= free) occurrences of n in X by E , or "by substituting E for n in X".

For the result of this substitution, an amazing number of different notations is in use. So far, we used informally

$$(X \text{ with } n := E)$$

(which can be read as "X with n replaced by E"). Another notation in common use is

$$X^n_E$$

but we shall not use it, for it has more than one disadvantage. It is typographically clumsy in that super- and subscripts invite smaller print (which invitation, when accepted recursively, is guaranteed to lead to unreadability —in particular in conjunction with a matrix printer!—); uneven spacing of the lines is another awkward consequence. Furthermore we have observed that in the case of simple E , such as in

$$X_m^n \quad ,$$

the absence of any visual aids to remind the reader what is substituted for what is definitely a shortcoming. Other notations we have seen are

$$X[n \leftarrow E] \quad , X(n := E) \quad , X(E/n) \quad , X(n \backslash E) \quad ,$$

the last two having been inspired by a metaphor of multiplication and division: the "denominator" disappears and the "numerator" comes in its place. They are linear and do give the reader some visual aid, and as such they are an improvement.

Yet, they do not fully satisfy us. The point is that this time we don't just need some representation of that boolean expression: we want to view that boolean expression —that predicate on the state space— as the result of applying a predicate transformer to predicate X , and we want our notation to reflect that view, because, this time, our interest is in that predicate transformer and its mathematical properties.

In view of our notational conventions adhered to so far, a postfix predicate transformer would be a notational anomaly. We therefore propose to write

$$(n := E).X \quad .$$

The advantage of the explicit full stop ($=$ period) for functional application is here that it is a visual reminder that the syntactic unit preceding it is a function to be applied to the syntactic unit following it.

Let now X and Y be two predicates on a state space of which n is a state variable. Because the replacement of n by E has to be done for *all* occurrences of n in the argument, substitution distributes by definition over all logical connectives, i.e.,

$$[(n := E).(X \equiv Y) \equiv (n := E).X \equiv (n := E).Y]$$
$$[(n := E).(X \lor Y) \equiv (n := E).X \lor (n := E).Y]$$
$$[(n := E).(X \land Y) \equiv (n := E).X \land (n := E).Y]$$
$$[(n := E).(\neg X) \equiv \neg(n := E).X]$$
$$[(n := E).(X \Rightarrow Y) \equiv (n := E).X \Rightarrow (n := E).Y] \quad .$$

Let now X be a function that, applied to some domain, yield a predicate on the state space; with a fresh dummy i ranging over that domain, we have by the definition of substitution

$$[(n := E).(\forall i :: X.i) \equiv (\forall i :: (n := E).(X.i))]$$
$$[(n := E).(\exists i :: X.i) \equiv (\exists i :: (n := E).(X.i))] \qquad ,$$

because all possible dependence on n is concentrated in the term. Alternatively, let V be a set of predicates on the state space; then the set V does *not* depend on n, only its elements possibly do, and we have with a fresh dummy X

$$[(n := E).(\forall X: X \in V: X) \equiv (\forall X: X \in V: (n := E).X)] \qquad ,$$
$$[(n := E).(\exists X: X \in V: X) \equiv (\exists X: X \in V: (n := E).X)] \qquad .$$

In other words: substitution is universally junctive.

Remark Probably the generally most respected way of representing the effect of substitution uses the λ-calculus: its notation would be

$$(\lambda n : X).E \qquad .$$

Here "$\lambda n :$" effectuates "functional abstraction": $(\lambda n : X)$ is the one and only function that yields X when it is applied to n; in order to effectuate the substitution by E, this function is applied to E.

The λ-calculus is certainly not without appeal. It has, for instance, the very clear rule that, while n may be a global variable of X, it is most definitely *not* a global variable of $(\lambda n : X)$. In this last expression, n has in fact the status of a dummy with its scope delineated by the surrounding parenthesis pairs; it has the status of a dummy in the sense that it may freely be replaced by a fresh variable, say n', i.e., (mixing notations!) we could rewrite $(\lambda n : X)$ as

$$(\lambda n' : X^n_{n'}) \qquad .$$

Yet we preferred to leave the λ-calculus alone because we don't need functional abstraction in isolation, but only in combination with a subsequent application. By presenting this combination "in a single package", so to speak, we have a syntactically more homogeneous manipulation and are spared all sorts of questions such as whether to define

$$[(\lambda n : (\forall X: X \in V: X)) \equiv (\forall X: X \in V: (\lambda n : X))] \qquad .$$

We would, however, like to warn the reader in this connection that our "packaged" function $(n := E)$ is —like most "neat ideas"— not without problems. The problems emerge as soon as we consider $(n := E)$ as an expression —there is no objection to considering expressions whose values are functions— and ask for its global variables. Certainly the global variables of E are global variables of $(n := E)$. But what about n? To simplify the discussion, let us take a constant for E, e.g., let us consider the predicate transformer $(n := 1)$. Here n is not a global variable because we are not allowed to substitute an arbitrary expression F for it: as predicate transformer $(F := 1)$ would be meaningless because textual substitution for a general expression cannot be defined in a useful way. So, n is not a global variable of $(n := 1)$. Yet n is an external handle of $(n := 1)$ in the sense that the substitutions $(n := 1)$ and $(m := 1)$ are different predicate transformers. Furthermore, note that —like quantification over n — application of $(n := 1)$ yields a result of which n is not a global variable.

With the introduction of $(n := E)$ as an expression, we have introduced externally significant variables of a different kind than global variables. Shall we call them "celestial variables"? Then $(n := n - m)$ would have the celestial variable n and the global variables n and m; we could even distinguish n's celestial occurrence from its global one. Once the celestial variable has been admitted, it requires a very strong mind to abstain from introducing the substitution that would transform the function $(n := n - m)$ into the function $(m := n - m)$.

(End of Remark.)

We close this chapter with some theorems about the substitution operator as predicate transformer. To pave the way we recall the principle of Leibniz —see (3, 2)—

$$[x = y] \Rightarrow [f.x = f.y]$$

and the one-point rule —see (5, 91)—

$$[(\forall x: [x = y]: f.x) \equiv f.y] \qquad .$$

In the case of punctual functions, we have —see (3, 6)— the stronger version of Leibniz:

(64) $[x = y \Rightarrow f.x = f.y]$

or, for boolean f, equivalently

$$[x = y \wedge f.x \equiv x = y \wedge f.y] \qquad .$$

Similarly, there is a stronger version of the one-point rule:

(65) $[(\forall x: x = y: f.x) \equiv f.y]$ for punctual f .

Proof We observe for any y and punctual f

$(\forall x:\ x = y:\ f.x)$

$=\quad \{(64)\}$

$(\forall x:\ x = y:\ f.y)$

$=\quad \{\text{pred. calc.}\}$

$(\forall x:\ x = y:\ false \vee f.y)$

$=\quad \{\ \vee\quad \text{distributes over}\ \forall\ \}$

$(\forall x:\ x = y:\ false) \vee f.y$

$=\quad \{\text{range}\ x\ \text{non-empty}\}$

$f.y\qquad .$

(*End of Proof.*)

In (65), the variables x and y stand for structures of the same type; x being a dummy, we may rewrite (65) as

$$[(\forall n:\ n = y:\ f.n) \equiv f.y]$$

even if n denotes one of the variables of the state space. (Due to the quantification, the left-hand side does not have n as global variable.) With n one of the variables of the state space, $f.n$ is just a given predicate; giving it a name, Fn say —chosen to remind us that it depends punctually on the otherwise anonymous but here explicitly named state variable n — , the corresponding way of denoting $f.y$ is $(n := y).Fn$. Thus we may rewrite

(66) $[(\forall n:\ n = y:\ Fn) \equiv (n := y).Fn]$

in which y is a global variable of both sides while n is not. We may instantiate it with $y := E$ *provided* n *is not a global variable of* E :

(67) $[(\forall n:\ n = E:\ Fn) \equiv (n := E).Fn]$.

Were we to instantiate, instead of with E , with En —i.e., an expression that has n as a global variable— then we would introduce in the left-hand side a classical example of clashing identifiers, viz., the global n of En and the dummy n of the universal quantification. The equally classical way out is to rename in (66) the dummy n by a fresh variable x

(68) $[(\forall x:\ x = y:\ (n := x).Fn) \equiv (n := y).Fn]$

and to instantiate the latter with $y := En$:

(69) $[(\forall x:\ x = En:\ (n := x).Fn) \equiv (n := En).Fn]$,

where x is a fresh variable.

Note that with the aid of (66) we can eliminate substitution from the left-hand side:

(70) $[(\forall x\colon\ x = En\colon\ (\forall n\colon\ n = x\colon\ Fn)) \equiv (n := En).Fn]$

where x is a fresh variable.

All this is a little bit painful. For the application of the substitution operator $(n := En)$, we need two nested quantifications, the inner one to introduce a fresh variable x and the outer one to eliminate it again. Here, the λ-calculus, which would have allowed us to write simply

        $(\lambda n\colon Fn).En$,

is at a definite advantage. Fortunately, we pay only a small price for this cumbersome transcription of the substitution operator, since later we shall use (70) only once.

CHAPTER 7

Semantics of straight-line programs

Program execution as change of machine state

This is a monograph about a theory of programming language semantics. Programming language definitions traditionally consist of two parts, called its "syntax" and its "semantics", respectively.

The syntax of a programming language defines which character sequences are programs in that programming language. Moreover, the syntax defines how programs written in that programming language are to be parsed. (Fortunately, syntaxes can, in turn, be defined in several distinct steps. A meaningful separation of concerns isolates, for instance, the consequences of the fact that, ultimately, each program is expressed as a linear string of characters from a finite alphabet. One step does not distinguish between programs and their parse trees and defines which parse trees belong to the language. Another step is exclusively concerned with how the parse trees belonging to the language are coded as linear strings of characters; this step settles details such as how scopes are delineated, whether there will be infix operators, etc. This monograph being about semantics, we shall not pursue the different aspects of syntax definition.)

The semantics of a programming language defines for each program written in that language "what that program means". (Here we have followed how the term semantics is used in linguistics.) The semantics refers to the execution of the programs, i.e., what would happen each time such a program is fed into an appropriate computer.

Remark We regret the terminology we used in the preceding paragraphs, but the terminology is by now so firmly established that there seems to be no point anymore in deviating from it. In retrospect, we —and not we alone— would have been much happier, had programming languages never been called "languages" (but, say, "program notations" instead). Admittedly, the metaphor has been helpful in the very beginning: the linguistic analogy gave a hint what aspects to distinguish and provided a terminology for identifying them. Our regret comes from the fact that, beyond that initial service, the linguistic analogy has been more confusing than helpful. (Is it far-fetched to suppose that the analogy has stood at the cradle of the faddish term "computer literacy"? We have seen the "argument" that computer literacy —whatever it may be: the author left that open— would be easier to acquire than (normal) literacy because programming languages are simpler than natural languages. In the same vein we have seen it "argued" that one should not strive for a programming language with a stable definition because the living languages are much more useful than the dead ones. And it is not only the layman that gets confused: the above two almost perfect examples of medieval reasoning have been displayed by two very well-known professors of computer science.)

The term "programming language" was first a symptom, but, once established, it became a driving force behind the habit of describing computing systems in anthropomorphic terminology. ("When this guy wants to speak to that guy···" in reference to two components of a computer network.) Having that habit is a severe handicap, and too many people suffer from it. As long as we don't shed it, computing science will remain immature. (*End of Remark*.)

Now, one aspect of doing science consists in choosing the dividing line between the relevant and the irrelevant. In the current case, the question is: what are we going to ignore and what are we going to take into account when referring to "what would happen each time such a program is fed into an appropriate computer"?

To begin with —and not surprisingly so— we are going to ignore all physical characteristics of the computer: its size, its weight, its price, its speed, its power consumption, its reliability, its maintenance contract, and its manufacturer, just to mention a few irrelevantia. Presumably the machine has a manual, but in that case we do not regard it as the manual's task to describe the machine, but as the machine's task to provide a —hopefully correct— physical model of the manual's contents. In case of a discrepancy, we blame the machine and not the manual: it is the abstract machine that matters, because that is the one we can think about.

Historical Note We found this understanding essential from the very first beginning of our cooperation. In the fifties, we were involved in the design, construction and method of use of a number of computers —the software was EWD's responsibility while CSS had responsibility for the hardware— . For each new machine, the writing of the manual was the very first thing that happened. These manuals were understood to state rights and obligations; to the extent they did, they almost acted like legal documents. Thus it was possible to complete the basic software and the prototype at the same moment. (*End of Historical Note.*)

Having decided that it is the thinkable machine that really matters, we go one step further and ignore all sorts of constraints that are hard to avoid in physical machines, such as limits on the size of store or on the length of words: in the machine we are willing to think about —also known as "the Good Lord's Machine"— any integer can be increased by 1 and there is no fixed limit on the depth of recursion.

Remark The thinkability of the Good Lord's Machine is the reason why, in order to be reasonably suitable for the implementation of "higher level programming languages", a fixed-wordlength machine should be equipped with dedicated hardware for overflow detection. A partial simulation of the Good Lord's Machine can be quite useful provided (i) the simulation can be made to be almost always successful, and (ii) the implementation signals if the simulation has failed. Hence, overflow detection is a very skew test, i.e., one that yields very little information per test. For such tests —parity check and real-time interrupt are other examples— dedicated hardware is needed to protect the implementor from the pressure to suppress the test. (*End of Remark.*)

To simplify matters further, we ignore all devices for input and output of information: the value of the input absorbed by a computation is deemed to be captured by the initial state of the machine —i.e., the state in which the machine starts the computation— , the value of the output is similarly deemed to be captured by the final state of the machine —i.e., the state in which the machine is left upon completion of the computation— .

Remark Our decision to ignore input and output commands is a conscious departure from the traditional computational model in which the output is produced as function of the input, commonly pictured as a box with an incoming arrow at its left and an outgoing arrow at its right. To someone grown up with that paradigm, the box without arrows —or, equivalently, programming without input and output commands— might seem unrealistically incomplete.

In this connection, we would like to point out that, with the inclusion of nondeterminacy, the direction of the input arrow has become somewhat arbitrary: we can replace an input file by an output file if we replace the input of a next value by a nondeterministic choice of a next value combined with the output of the value chosen. What used to be a relation between input and output is thus transformed into a relation between two outputs. Similarly we can replace an output file by an input file miraculously containing the answers if we replace the output of a value by a comparison with the next value read.

These observations strongly support our decision not to give input and output a special status in our considerations. The reader worried by the above should remember that the notion of causality has no useful rôle to play in the inanimate world; if he believes it does, he is probably the victim of one or two anthropomorphic metaphors. (*End of Remark.*)

Our next step is a little bit more radical. In the early days of automatic computing we were very preoccupied with the machines, whose construction —to put it mildly— stretched the electronic technology of the day. As a result, it was viewed as the task of the programs to instruct our machines, and the final state was not only viewed but actually defined as the result of the computation evoked under control of the program. In those days, the semantics of programming languages was defined in terms of the computational steps that would take place during the executions of the programs, the final state reflecting the cumulative effect of all the steps of a computation. In order to distinguish it from more modern ways, the approach just sketched is now called "an operational definition" of the semantics.

Nowadays, we are no longer so preoccupied with the machines. We are more interested in our programs and view it as the task of the machines to execute our programs, i.e., to bring themselves, for given initial state and given program, into a corresponding final state. We are not so interested any more in what happens during the computation, provided it establishes the desired nett effect. Accordingly, we shall no longer define the semantics via the detour of the nett effect of the computations a program may evoke; instead we shall define more directly how, for any program in the language, initial and final states are connected. In order to distinguish it from operational definitions, the latter is called "a postulational definition" of the semantics.

We adopt a postulational method because its technical advantages, as compared with operational methods, are so overwhelming. (It serves as the basis for a calculus to derive programs from their specifications; most, if not all, of the functions to be manipulated are total; it caters for the painless inclusion of nondeterminacy.)

The only things of interest that remain are the initial state, the final state, and how a program defines a connection between them. Since computations no longer enter the picture, we can forget about machines and computational models. The postulational method allows us to treat programs as uninterpreted formulae, i.e., mathematical objects in their own right, that we can deal with while ignoring the fact that they are interpretable as executable code.

Remark Postulational methods relegate what used to be considered the subject matter to the secondary rôle of (ignorable) model in terms of which the new formalism could be interpreted. It is a time-honoured tradition that their introduction begins by evoking sometimes bitter resentment from those that are so attached to their familiar patterns of reasoning that they experience the reduced rôle of their cherished model as an insufferable loss. But, to quote E. T. Bell: "Experience shows that the only loss is the denial of the privilege of making avoidable mistakes in reasoning.". In programming it is the operational model of computation that invites the "avoidable mistakes in reasoning".

We mention one further advantage of the postulational method that emerged in programming. In order to reason in a trustworthy manner about abstract programs, one has to know how to cope with unbounded nondeterminacy, something for which no physically realistic model of computation exists. For the operationally inclined this has proved to be a severe hurdle, whereas the postulational method takes it without noticing that it could be a hurdle. (*End of Remark.*)

The semantics of a program

One way of trying to capture the semantics of a given program S is to consider the final state in which S terminates as a function of the initial state in which S has been started. This way has been followed —even extensively so— but it has a few serious disadvantages.

One disadvantage is that there may be initial states for which S , when started in them, fails to terminate, i.e., the final state, considered as a function of the initial state, is not defined for all points of the state space. We then have two options. Either we pay the price of dealing with what are called "partial functions" —i.e., functions defined on a smaller domain than the one under consideration— or we make them total functions by extending the range with a special value —usually called "bottom"— that can be interpreted as "stuck in an eternal computation". Both have their disadvantages. The first method introduces functions that cannot be freely applied to an argument that is otherwise of the appropriate type, the second method buys the advantage of dealing with total functions at the price of destroying the homogeneity of the state space.

The other disadvantage is more serious: the functional approach is really geared to what are now called "deterministic programs", i.e., programs for which the ensuing computation —whether finite or not— is fully determined by occasionally tossing a coin —with enough sides— simply don't fit nicely reasons it is desirable to include nondeterministic programs in our considerations as well, i.e., programs in which the ensuing computation is only partly determined by the initial state. Such programs, which could be implemented by occasionally tossing a coin —with enough sides—, simply don't fit nicely in the functional framework. Of course people have tried to squeeze them into it: one effort was to treat the behaviour of the coin as some sort of hidden input, another one introduced functions from states to subsets of states. The cleanest approach along these lines is probably to capture the semantics of a program as a relation between initial and final state; it has at least the advantage that the relational calculus as such is available. In this context, experience with the relational calculus has not been too favourable. The transition to the relational calculus does almost suffice for the elimination of "bottom", but not quite! Moreover, the relational calculus, which treats the two arguments of a relation on the same footing, does not, by itself, reflect the asymmetry between initial and final state. These are admittedly only tentative explanations for the not-too-fortunate experience with the relational calculus. Other possible explanations are that no one trying to apply the relational calculus in this area mastered it well enough, or that the relational calculus needs a few notational revisions before it can be considered a workable tool.

After the above, we trust that the reader understands our preference for a definition of semantics
(i)   that uses functions, rather than relations,
(ii)   of which the functions are total rather than partial, i.e., one that allows us to dispense with "bottom", and
(iii) that encompasses nondeterminacy.

To begin with, we look for a helpful classification of computations. A major dichotomy is into terminating and eternal computations: a terminating computation has an initial state and a final state, whereas an eternal computation has only an initial state.

For the purpose of semantic definition, the dichotomy into terminating and eternal computations is too coarse because it treats all points of the state space on the same footing. The general way of distinguishing points in state space is the introduction of some predicate X ; a predicate X defines a dichotomy of the state space in the sense that each state satisfies either X or $\neg X$.

Using the latter dichotomy to refine the former, we have two options: we can relate predicate X to the initial state or to the final state. In the first case we would introduce four classes of computations according to "terminating/ eternal" and "initially X /initially $\neg X$". In the second case, where we relate predicate X to the final state, this only yields a dichotomy of the terminating computations and hence we would introduce only three classes: eternal/ finally X /finally $\neg X$.

We have to choose between the two options. Fortunately, some reflection tells us that the second one can be expected to lead to the more convenient concepts. In the second classification, "eternal/finally X /finally $\neg X$", we have postponed taking the initial state into account, but that will be no problem because each computation of each class has an initial state. In the first classification, into four classes, we have postponed taking the final state into account, and when we try to do so, the trouble starts for not all computations of all classes do have a final state. In summary, we propose to partition the computations into the following —indeed mutually exclusive— classes:

"eternal"   —all computations that fail to terminate

"finally X"   —all computations terminating in a final state
                       satisfying X

"finally $\neg X$"   —all computations terminating in a final state
                       satisfying $\neg X$.

Note that —thanks to the Excluded Middle— this classification is also exhaustive: each computation falls into one of the three classes.

So much for the classification of computations with respect to the predicate X . We now take the view that we know everything that is to be known about the semantics of a program S if we know, for any predicate X and for any initial state, computations from which of the three classes are possible when S is started in that initial state.

To each class then corresponds a dichotomy of the initial state space, viz., whether in such a state as starting point an execution of that class is possible or not. We characterize —as always— these dichotomies by predicates on the state space. For given S and X we propose the following three predicates (whose nomenclature will be explained later). The first of three predicates is independent of X .

$wp.S.true$:   holds precisely in those initial states for which no
                    computation under control of S belongs to the class
                    "eternal"

$wlp.S.X$:   holds precisely in those initial states for which no
                    computation under control of S belongs to the class
                    "finally $\neg X$"

wlp.S.($\neg X$)   :   holds precisely in those initial states for which no computation under control of S belongs to the class "finally X" .

Because each computation belongs by definition to exactly one of the three classes, an alternative phrasing is

wp.S.true   :   holds precisely in those initial states for which each computation under control of S belongs either to the class "finally X" or to the class "finally $\neg X$"

wlp.S.X   :   holds precisely in those initial states for which each computation under control of S belongs either to the class "eternal" or to the class "finally X"

wlp.S.($\neg X$)   :   holds precisely in those initial states for which each computation under control of S belongs either to the class "eternal" or to the class "finally $\neg X$" .

Above denotations of predicates on the initial states —such as "*wp.S.true*" and "*wlp.S.X*" — are not just names: the full stops (= periods) are the left-associative full stops (= periods) denoting function application —we could equally well have written "(*wp.S*).*true*" and "(*wlp.S*). X" — . That is, *wp.S* and *wlp.S* , being functions from predicates to predicates, emerge as predicate transformers. (The first one has, so far, only been applied to the argument *true* , but that will be remedied shortly.) In short, we treat *wp* and *wlp* as higher-order functions.

As we said, we consider the semantics of program S fully characterized by the knowledge of the predicates *wp.S.true* and *wlp.S.X* for all X , i.e., the knowledge of the predicate *wp.S.true* and the predicate transformer *wlp.S* .

Not all predicate transformers can meaningfully be interpreted as a *wlp.S* for some program S . To begin with we observe for any (possibly empty) bag V of predicates and any *terminating* computation C

$$(C \text{ belongs to the class "finally } (\forall X: X \in V: X)\text{"}) \equiv$$
$$(\forall X: X \in V: C \text{ belongs to the class "finally } X\text{"}) \qquad .$$

Hence, for any C , the above equivalence holds or C belongs to the class "eternal". Disjunction distributing over equivalence and over universal quantification, we deduce for any V and C

$$(C \text{ belongs either to the class "eternal" or to the class "finally}$$
$$(\forall X: X \in V: X)\text{"}) \equiv$$
$$(\forall X: X \in V: C \text{ belongs either to the class "eternal" or to the}$$
$$\text{class "finally } X\text{"}) \qquad .$$

From this observation we conclude that our interpretation requires that for each program S

(0) $wlp.S$ is universally conjunctive   ;

because we shall need it in a moment, we mention universal conjunctivity's special consequence: we have for each program S

(1) $[wlp.S.true]$.

And now the time has come to introduce, in terms of the predicate $wp.S.true$ and the predicate transformer $wlp.S$, the second predicate transformer we associate with program S . It is the predicate transformer $wp.S$ given by

(2) $[wp.S.X \equiv wp.S.true \wedge wlp.S.X]$ for all X .

(Note that, thanks to (1), (2) with $X := true$ does not lead to a conflict: we were justified in naming the first of the three characteristic predicates $wp.S.true$.)

From the interpretation of the conjuncts in the right-hand side of (2) we derive the interpretation of $wp.S.X$:

$wp.S.X$: holds precisely in those initial states for which each computation under control of S belongs to the class "finally X ".

The names wp and wlp are derived from "weakest precondition" and "weakest liberal precondition", respectively: $wp.S.X$ is "the weakest precondition under which S is guaranteed to establish the postcondition X ", $wlp.S.X$ is "the weakest precondition under which S is guaranteed to establish the postcondition X if the computation terminates". In the jargon: $wlp.S$ is concerned with "the partial correctness of S " (i.e., apart from possible failure to terminate), whereas $wp.S$ is concerned with "the total correctness of S " (i.e., termination included).

Remark When C. A. R. Hoare introduced what became known as "Hoare triples" and got the format

(3) $\{P\}\ S\ \{Q\}$,

he was concerned with partial correctness: (3) was short for "when started in an initial state satisfying P , S will, if the computation ends, end in a final state satisfying Q ". The relation with our notation is that $wlp.S.X$ is the weakest solution of the equation

$$Y: (\{Y\}\ S\ \{X\})   .$$

(Note that in this equation $wlp.S.X$ emerges as extreme solution without being a "fixpoint", i.e., not a solution of an equation of the form $Y: [Y \equiv f.Y]$. Sometimes people seem to forget that such extreme solutions do exist.)

Hoare's triples and inference rules were designed to formalize reasoning about (partial) correctness of programs. By switching to predicate transformers the triples have been eliminated and the inference rules, if not subsumed in the predicate calculus, have been replaced by definitions or theorems. (*End of Remark.*)

Of the two, $wp.S$ and $wlp.S$, the latter is the more fundamental one —there is no way of defining $wlp.S$ in terms of $wp.S$ — ; it is also the one with the nicer properties (which probably reflects that it is easier to reason about partial correctness than about total correctness). We include $wp.S$ in our considerations because in program design it is often the more useful one. Besides that, it has some interesting properties.

We observe that in our classification of computations the class "finally *false*" is empty (for lack of final states satisfying *false*). Hence —see our interpretation of $wp.S.X$ — we have the interpretation

$wp.S.false$: holds precisely in those initial states for which no computation under control of S exists.

As long as we take the position that each program S can be started in any initial state, i.e., that to any initial state at least some computation corresponds, then there are no initial states for which no computation under control of S exists. That is, we have for each program S

(4) $[wp.S.false \equiv false]$.

Because a computation in the class "finally *false*" would establish a miracle, one of us (EWD) dubbed (4) "The Law of the Excluded Miracle".

Remark For this catchy name EWD has been severely chided (by Greg Nelson) and rightly so. The catchy name, together with the fact that it had been called a "Law", erected a considerable barrier for the conception of program components S for which $wp.S.false$ would hold in some states, viz., those initial states in which starting S would be simply "inappropriate" or "impossible". The introduction of the perfectly sound concept of such partial programs —"partial" in the sense that in some initial states there is no computation to take place under control of the program— has been delayed by EWD's nomenclature. For which he offers his apologies. In this little monograph we shall confine ourselves to "total" programs satisfying (4). (*End of Remark.*)

From the interpretation

$wlp.S.(\neg X)$: holds precisely in those initial states for which no computation under control of S belongs to the class "finally X"

we deduce the interpretation

$\neg wlp.S.(\neg X)$:   holds precisely in those initial states for which there exists a computation under control of S that belongs to the class "finally X".

Whereas $wp.S.X$ characterizes the initial states where "finally X" is unavoidable, $\neg wlp.S.(\neg X)$ characterizes those states for which "finally X" is merely possible. Hence we should have

(5) $[wp.S.X \Rightarrow \neg wlp.S.(\neg X)]$ for all X .

Indeed we observe for any S and X

$\quad [wp.S.X \Rightarrow \neg wlp.S.(\neg X)]$
$= \quad \{\text{pred. calc.}\}$
$\quad [\neg(wp.S.X \wedge wlp.S.(\neg X))]$
$= \quad \{(2)\}$
$\quad [\neg(wp.S.true \wedge wlp.S.X \wedge wlp.S.(\neg X))]$
$= \quad \{\ wlp.S \text{ is universally conjunctive}\}$
$\quad [\neg(wp.S.true \wedge wlp.S.false)]$
$= \quad \{(2)\}$
$\quad [\neg wp.S.false]$
$= \quad \{(4)\}$
$\quad true$.

Note The above calculation would have been two steps shorter, had we used the theorem that for any S , X , Y

(6) $[wp.S.X \wedge wlp.S.Y \equiv wp.S.(X \wedge Y)]$,

which the reader may prove. (*End of Note.*)

A program being deterministic means that, conversely, what is possible is also unavoidable, i.e., S being deterministic means

$\qquad [wp.S.X \Leftarrow \neg wlp.S.(\neg X)]$ for all X .

Combining this result with (5) and observing that by the definition of the conjugate

$\qquad [\neg wlp.S.(\neg X) \equiv (wlp.S)^*.X]$,

we arrive at the definition

(7)   (S is deterministic) $\equiv$
            ($wp.S$ and $wlp.S$ are each other's conjugate)   .

In this booklet the concept of determinacy does not play a very important rôle. The above definition was, in fact, only introduced at a fairly late stage. We think it worthwhile to note that, as a definition, it is very nice and, by being so, gives an encouraging indication that we have introduced appropriate concepts. This was a further reason for its inclusion.

And now we finish our operational considerations and will proceed to define the semantics of programs S by defining the corresponding predicate transformers $wlp.S$ and $wp.S$. The *only* connection with the above will be a purely formal one:

(i) for every $wlp.S$ we define, we shall honour the obligation to show that it meets requirement

R0: $wlp.S$ is universally conjunctive   ;

(ii) for every $wp.S$ we define, we shall honour the obligation to show that it meets requirement

R1: $[wp.S.false \equiv false]$.

We close this section with two simple theorems.

(8) *Theorem* For any S , $wp.S$ is positively conjunctive.

Proof On account of (2) —the definition of $wp.S$ — and (6, 41), $wp.S$ enjoys each type of conjunctivity shared by the constant function and $wlp.S$. The constant function —see (6, 37)— is positively conjunctive, $wlp.S$ —see R0 — is universally so. From the combination of these two observations, the theorem follows.

(*End of Proof.*)

(9) *Theorem* For any deterministic program S , $wp.S$ is universally disjunctive, and $wlp.S$ positively so.

Proof We observe for any deterministic S

    the disjunctivity type of $wp.S$
$=$ $\{(7); \ S$ is deterministic$\}$
    the disjunctivity type of $(wlp.S)^*$
$=$ $\{$properties of conjugate$\}$
    the conjunctivity type of $wlp.S$
$=$ $\{$ R0 $\}$
    universal   .

The second half of the proof is left to the reader.

(*End of Proof.*)

The semantics of a programming language

So far, we argued that we can consider the semantics of S defined if the predicate transformers $wlp.S$ and $wp.S$ are defined; the definition of such predicate transformers and the study of their properties will be the subject of the remainder of this chapter.

The syntax of a programming language defines the set of all programs that are writable in it. The semantics of a programming language has to define the semantics of each writable program, i.e., has to define for each writable program S the predicate transformers $wlp.S$ and $wp.S$. In other words, the definition of the semantics of a programming language boils down to a definition of the functions wlp and wp , which are functions from writable programs to predicate transformers. Their domain, i.e., the writable programs, being recursively defined by the grammar, we shall define wlp and wp recursively over the grammar. For each statement S we shall list in order

(i)   the predicate transformer $wlp.S$
(ii)   the predicate $wp.S.true$
(iii)   the predicate transformer $wp.S$,

where (i) and (ii) should be viewed as definitions from which —and the reader is supposed to verify this himself— (iii) then follows according to (2). Next we shall discharge R0 and R1 . Finally, we may discuss some of its properties. We postpone the introduction of compound statements until after the introduction of a bunch of basic statements.

havoc

(10) $[wlp.havoc.X \equiv [X]]$ for all X

(11) $[wp.havoc.true \equiv true]$

(12) $[wp.havoc.X \equiv [X]]$ for all X .

Because $wlp.havoc$ is the "everywhere" operator and —see (6, 40)— the "everywhere" operator is universally conjunctive, requirement R0 is met.

In order to see that requirement R1 is met, we observe

$wp.havoc.false$
= $\{$(12) with $X := false$ $\}$
$[false]$
= $\{$pred. calc.$\}$
$false$.

Comparing (10) and (12), we see that *wlp.havoc* and *wp.havoc* are the same predicate transformer; hence *wp.havoc* is universally conjunctive.

To see that *wlp.havoc* is not finitely disjunctive we observe for any Y that is neither *true* nor *false*

$$wlp.havoc.(Y \vee \neg Y)$$
$$= \quad \{(10) \text{ with } X := Y \vee \neg Y \}$$
$$[Y \vee \neg Y]$$
$$= \quad \{\text{pred. calc.}\}$$
$$true$$

whereas

$$wlp.havoc.Y \vee wlp.havoc.(\neg Y)$$
$$= \quad \{(10) \text{ with } X := Y \text{ and } X := \neg Y \}$$
$$[Y] \vee [\neg Y]$$
$$= \quad \{ \quad Y \text{ is neither } true \text{ nor } false \}$$
$$false \qquad .$$

To see that *wlp.havoc* is not *or*-continuous either, we observe for some state space with natural state variable n —and dummy i understood to be natural—

$$wlp.havoc.(\exists i:: \quad n \leqslant i)$$
$$= \quad \{(10) \text{ with } X := (\exists i:: \quad n \leqslant i)\}$$
$$[(\exists i:: \quad n \leqslant i)]$$
$$= \quad \{\text{consider } i = n \}$$
$$true$$

whereas

$$(\exists i:: \quad wlp.havoc.(n \leqslant i))$$
$$= \quad \{(10) \text{ with } X := n \leqslant i \}$$
$$(\exists i:: \quad [n \leqslant i])$$
$$= \quad \{\text{consider } n = i + 1 \}$$
$$(\exists i:: \quad false)$$
$$= \quad \{\text{pred. calc.}\}$$
$$false \qquad .$$

Hence, apart from monotonicity, *wlp.havoc* is not disjunctive and neither is *wp.havoc* . Consequently, —see (9)— *havoc* is not deterministic.

In fact, *havoc* is almost as nondeterministic as possible. Operationally interpreted, the only thing we know about an execution of *havoc* is that it terminates; upon its termination the machine may be in any state, i.e., all variables spanning the state space may have been set to unpredictable, unrelated values. For that reason, most commercially available programming languages do not include —as least not intentionally— *havoc* in their repertoire. Yet it is good to know *havoc* , as it may be valuable in the design of counter-examples.

Note In an infinite state space, the nondeterminacy of *havoc* is, in the jargon, "unbounded". The weakening sequence $(n \leqslant i)$ for $(0 \leqslant i)$ is the standard tool for showing that, for an S of unbounded nondeterminacy, *or*-continuity of *wlp.S* is excluded. (*End of Note.*)

abort

(13) $[wlp.abort.X \equiv true]$ for all X

(14) $[wp.abort.true \equiv false]$

(15) $[wp.abort.X \equiv false]$ for all X .

Because *wlp.abort* is the constant function *true* and —see (6, 39)— the constant function *true* is universally conjunctive, requirement R0 is met.

Because $[wp.abort.false \equiv false]$, requirement R1 is met.

We observe for any X

$(wlp.abort)^*.X$
$=$ {definition of conjugate}
$\neg wlp.abort.(\neg X)$
$=$ {(13) with $X := \neg X$ }
$\neg true$
$=$ {pred. calc.}
$false$
$=$ {(15)}
$wp.abort.X$,

i.e., *abort* is deterministic; hence *wlp.abort* is positively disjunctive and *wp.abort* is universally disjunctive.

The operational interpretation of *abort* is that for all initial states its execution fails to terminate.

skip

(16) $[wlp.skip.X \equiv X]$ for all X

(17) $[wp.skip.true \equiv true]$

(18) $[wp.skip.X \equiv X]$ for all X .

Because *wlp.skip* is the identity function and —see (6, 36)— the identity function is universally conjunctive, requirement R0 is met.

Because $[wp.skip.false \equiv false]$, requirement R1 is met.

Because in fact the identity function is universally junctive and its own conjugate, *wlp.skip* and *wp.skip* are both universally junctive and *skip* is deterministic.

The operational interpretation of *skip* is that its execution, which is guaranteed to terminate, leaves the values of all variables unchanged.

"y := E"

Remark   We have chosen to delineate statements given by a piece of text rather than just an identifier by a pair of quotation marks, the parentheses not being available because $(y := E)$ is used to denote the substitution operator. (*End of Remark.*)

(19) $[wlp."y := E".X \equiv (y := E).X]$ for all X

(20) $[wp."y := E".true \equiv true]$

(21) $[wp."y := E".X \equiv (y := E).X]$ for all X .

Because $wlp."y := E"$ is a substitution and substitution is universally conjunctive, requirement R0 is met.

Because $[wp."y := E".false \equiv false]$, requirement R1 is met.

Because substitution as predicate transformer is, in fact, universally junctive and its own conjugate, $wlp."y := E"$ and $wp."y := E"$ are both universally junctive and "$y := E$" is deterministic.

The statement "$y := E$" is known as "the assignment statement". Its operational interpretation is that its execution, which is guaranteed to terminate, leaves the values of all variables, except y , unchanged, whereas the final value of y equals the initial value of E .

Remark We will not complicate the semantics of the assignment statement by admitting for E partial or multivalued expressions, such as p/q or $\pm p$, respectively, generalizations that can be viewed as a misuse of the functional notation. A major charm of defining program semantics in terms of $wlp.S$ and $wp.S$ is that, even in the presence of nontermination or nondeterminacy, these are (total and unique) functions of the postcondition. We won't allow the expression in the assignment statement to destroy this.

We shall make use of the traditional way of defining the value of expressions as the one and only root of an equation

(22) $x: [B.x]$,

where operands in the expression —or arguments of the function— occur as parameters in B , like defining $p - q$ as the root of $x: [p = q + x]$. We shall, in fact, define predicate transformers in that way, but shall carefully resist the temptation of introducing such B's that (22) may have more than one solution or no solution at all. (*End of Remark.*)

"S0;S1"

(23) $[wlp.\text{``}S0;S1\text{''}.X \equiv wlp.S0.(wlp.S1.X)]$ for all X

(24) $[wp.\text{``}S0;S1\text{''}.true \equiv wp.S0.(wp.S1.true)]$

(25) $[wp.\text{``}S0;S1\text{''}.X \equiv wp.S0.(wp.S1.X)]$ for all X .

This time we are not defining a basic statement: we define a new statement "S0;S1" in terms of the two statements $S0$ and $S1$. Accordingly, the predicate transformers characterizing the semantics of "S0;S1" are expressed in terms of the predicate transformers characterizing the semantics of $S0$ and $S1$. In particular, we shall prove the basic properties of "S0;S1" under the assumption that both $S0$ and $S1$ meet our standard requirements R0 and R1 .

In deviation from our earlier practice, we shall this time show that the conjunction of the right-hand sides of (23) and (24) yields that of (25). We observe for any X

$wp.S0.(wp.S1.true) \wedge wlp.S0.(wlp.S1.X)$
$=$ {(6) with S , X , $Y := S0$, $wp.S1.true$, $wlp.\ S1.X$ }
$wp.S0.(wp.S1.true \wedge wlp.S1.X)$
$=$ {(2) with $S := S1$ }
$wp.S0.(wp.S1.X)$.

Remark Note that the proof of (6) requires $wlp.S$ to be finitely conjunctive. (*End of Remark.*)

Because $S0$ and $S1$ meet requirement R0 , $wlp.S0$ and $wlp.S1$ are universally conjunctive and so is therefore —on account of (6, 45)— their functional composition, which —see (23)— is $wlp.``S0;S1"$ , i.e., $``S0;S1"$ meets requirement R0 .

To see that $``S0;S1"$ meets requirement R1 we observe

$wp.``S0;S1".false$

$=$ $\{(25)$ with $X := false$ $\}$

$wp.S0.(wp.S1.false)$

$=$ $\{$ $S1$ meets R1 $\}$

$wp.S0.false$

$=$ $\{$ $S0$ meets R1 $\}$

$false$.

Because —see (6, 45)— functional composition is junctivity preserving, $wlp.``S0;S1"$ enjoys each junctivity property shared by $wlp.S0$ and $wlp.S1$; similarly, $wp.``S0;S1"$ enjoys each junctivity property shared by $wp.S0$ and $wp.S1$.

Likewise, $``S0;S1"$ is deterministic if $S0$ and $S1$ are. We observe for any deterministic statements $S0$ and $S1$

$(wlp.``S0;S1")^*$

$=$ $\{(23)\}$

$((wlp.S0) \circ (wlp.S1))^*$

$=$ $\{$conjugate distributes over functional composition$\}$

$(wlp.S0)^* \circ (wlp.S1)^*$

$=$ $\{$ $S0$ and $S1$ are deterministic$\}$

$(wp.S0) \circ (wp.S1)$

$=$ $\{(25)\}$

$wp.``S0;S1"$.

Furthermore, we draw attention to the fact that, as infix operator on the domain of statements, the semicolon is associative because functional composition is associative, i.e.,

$$``(S0;S1);S2" = ``S0;(S1;S2)"   ;$$

we therefore can (and do) write just $``S0;S1;S2"$.
(The above equality means semantic equality:

$$S0 = S1$$

is short for

$$wlp.S0 = wlp.S1 \ \wedge \ wp.S0 = wp.S1$$

or, even more explicitly,

$$(\forall X :: [wlp.S0.X \equiv wlp.S1.X] \wedge [wp.S0.X \equiv wp.S1.X]) \quad .)$$

Note that the semicolon is neither symmetric nor idempotent.

Remark By defining for any $S0$ and $S1$ the semantics of "$S0;S1$" , this little section essentially defines the semantics of the semicolon as composition operator on statements. Whenever a way is given of composing a whole out of parts, that definition primarily describes how the relevant properties of the whole are determined by the relevant properties of the parts. If, for a proposed composition operator, that latter dependence turns out to be ugly, it is time to reconsider whether the composition operator proposed is really the one we want to work with. The semicolon passes the test of niceness with flying colours. (*End of Remark.*)

IF

The statement *IF* , standing for the alternative construct, is traditionally written as a list of "guarded commands" surrounded by the parenthesis pair "*if fi*" . A fat bar "[]" separates guarded commands in the list. A guarded command is of the form

$$\langle \text{boolean expression} \rangle \to \langle \text{statement} \rangle \quad .$$

Typical specimina of *IF* are of the form

if fi (= *abort*, see (27) and (29))
if $B.0 \to S.0$ *fi*
if $B.0 \to S.0$ [] $B.1 \to S.1$ *fi*   ,
if $B.0 \to S.0$ [] $B.1 \to S.1$ [] $B.2 \to S.2$ *fi*   ,
etc.

In the above, we have used functional notation so as to be able to refer to $B.i$ and $S.i$; in the following, the constraint on the range of i is left understood.

With the abbreviation BB given by

(26) $[BB \equiv (\exists i :: B.i)]$,

the semantics of *IF* satisfies

(27) $[wlp.IF.X \equiv (\forall i: B.i: wlp.(S.i).X)]$ for any X

(28) $[wp.IF.true \equiv BB \wedge (\forall i: B.i: wp.(S.i).true)]$

(29) $[wp.IF.X \equiv BB \wedge (\forall i: B.i: wp.(S.i).X)]$ for any X .

For the sake of completeness we shall demonstrate that the last line follows from the preceding two by observing for any IF and X

$$wp.IF.X$$
$= \quad \{(2) \text{ with } S := IF \ \}$
$$wp.IF.true \wedge wlp.IF.X$$
$= \quad \{(28) \text{ and } (27)\}$
$$BB \wedge (\forall i: \ B.i: \ wp.(S.i).true) \wedge (\forall i: \ B.i: \ wlp.(S.i).X)$$
$= \quad \{\text{pred. calc.}\}$
$$BB \wedge (\forall i: \ B.i: \ wp.(S.i).true \wedge wlp.(S.i).X)$$
$= \quad \{(2) \text{ with } S := S.i \ \}$
$$BB \wedge (\forall i: \ B.i: \ wp.(S.i).X) \qquad .$$

In order to convince ourselves that IF meets requirement R0 under the assumption that all the $S.i$ do, we rewrite $wlp.IF.X$ by trading as

$$(\forall i:: \ \neg B.i \vee wlp.(S.i).X) \qquad ;$$

the terms being universally conjunctive functions of X (on account of (6, 42)), we conclude on account of (6, 41) that $wlp.IF$ is universally conjunctive, i.e., that IF meets requirement R0 .

In order to convince ourselves that IF meets requirement R1 under the assumption that all the $S.i$ do we observe

$$wp.IF.false$$
$= \quad \{(29) \text{ with } X := false \ \}$
$$BB \wedge (\forall i: \ B.i: \ wp.(S.i).false)$$
$= \quad \{ \ S.i \text{ meets R1} \ \}$
$$BB \wedge (\forall i: \ B.i: \ false)$$
$= \quad \{(26)\}$
$$BB \wedge \neg BB$$
$= \quad \{\text{pred. calc.}\}$
$$false \qquad .$$

By (8), $wp.IF$ is positively conjunctive. Let us now turn our attention to its disjunctivity properties. By definition (2) of $wp.S.X$ and the dual of (6, 42), $wp.S$ enjoys all the disjunctivity properties of $wlp.S$, and we can therefore focus our attention on the disjunctivity properties of $wlp.IF$.

Let us rewrite (27) —by trading— again as

$$[wlp.IF.X \equiv (\forall i:: \ \neg B.i \vee wlp.(S.i).X)] \qquad .$$

Because, in general, the constant function $\neg B.i$ is only positively disjunctive,

$$\neg B.i \vee wlp.(S.i).X$$

is —on account of the dual of $(6, 41)$— as disjunctive as $wlp.(S.i)$, except for universal disjunctivity, in particular, *or*-continuous if $wlp.(S.i)$ is. On account of the dual of $(6, 43)$ we therefore have

(30) *Theorem* Predicate transformer $wlp.IF$ is *or*-continuous if all $wlp.(S.i)$ are *or*-continuous and the range of i is finite.

Remark The restriction to a finite range for i is essential. Consider, for a state space consisting of the natural variable n , the statement UN —short for "Unbounded Nondeterminacy"— given by

$$UN = \textbf{if}\,(\square i\colon\ 0 \leqslant i\colon\ true \to n := i)\,\textbf{fi}$$

(also known as "assign to n an arbitrary natural number"). Because all the guards —i.e., the $B.i$ — are *true* , and because $wlp."n := i"$ and $wp."n := i"$ are the same predicate transformer, $wlp.UN$ and $wp.UN$ are the same predicate transformer. According to (27) and (19)

(31) $[wlp.UN.X \equiv (\forall i\colon\ 0 \leqslant i\colon\ (n := i).X)]$.

In order to show that $wlp.UN$ is not *or*-continuous, we show for the weakening sequence $(j \geqslant n)$ for $j \geqslant 0$ that

$$[(\exists j\colon\ 0 \leqslant j\colon\ wlp.UN.(j \geqslant n)) \not\equiv wlp.UN.(\exists j\colon\ 0 \leqslant j\colon\ j \geqslant n)]\qquad.$$

To this end we observe

$\quad(\exists j\colon\ 0 \leqslant j\colon\ wlp.UN.(j \geqslant n))$
$=\quad\{(31)\ \text{with}\ X := j \geqslant n\ \}$
$\quad(\exists j\colon\ 0 \leqslant j\colon\ (\forall i\colon\ 0 \leqslant i\colon\ (n := i).(j \geqslant n)))$
$=\quad\{\text{def. of substitution}\ n := i\ \}$
$\quad(\exists j\colon\ 0 \leqslant j\colon\ (\forall i\colon\ 0 \leqslant i\colon\ j \geqslant i))$
$=\quad\{\text{consider}\ i = j + 1\ \}$
$\quad(\exists j\colon\ 0 \leqslant j\colon\ false)$
$=\quad\{\text{pred. calc.}\}$
$\quad false$

whereas

$\quad wlp.UN.(\exists j\colon\ 0 \leqslant j\colon\ j \geqslant n)$
$=\quad\{\text{consider}\ j = n\ \}$
$\quad wlp.UN.true$
$=\quad\{(1)\}$
$\quad true\qquad.$

Combination of these two observations shows that UN is not *or*-continuous. (*End of Remark.*)

In IF , each "$B.i \rightarrow S.i$" is called a "guarded command" and the $B.i$ is called "its guard". Showed the preceding remark that an infinite number of guarded commands may destroy *or*-continuity, the following example shows that a finite number of guarded commands may destroy finite disjunctivity. Consider S given by

$$S = \textbf{if } true \rightarrow n := 0 \; \square \; true \rightarrow n := 1 \textbf{ fi}$$

The reader is invited to verify that

$$[wlp.S.(n = 0 \lor n = 1) \neq wlp.S.(n = 0) \lor wlp.S.(n = 1)]$$

Because, for deterministic S , $wp.S$ is universally disjunctive and $wlp.S$ positively so, the observed losses of disjunctivity imply loss of determinacy. Let us therefore investigate under what circumstance we can conclude that IF is deterministic, i.e., that we have for all X

(32) $[wp.IF.X \equiv (wlp.IF)^*.X]$.

To begin with, we make two observations. Firstly, that (32) amounts to an equivalence between a universal quantification/conjunction and an existential quantification/disjunction, and, secondly, that there is every reason to suspect the presence of nondisjoint guards as the origin of the nondeterminacy. Combining the two observations we propose to investigate how we can relate the two differently quantified expressions under the assumption of disjoint guards.

To this end we first present two preliminary, general theorems.

(33) *Theorem* We have for any B and R

$$[(\forall i,j: \; B.i \land B.j: \; i = j) \Rightarrow$$
$$((\exists i: \; B.i: \; R.i) \Rightarrow (\forall j: \; B.j: \; R.j))]$$.

Proof We observe for any B and R

$(\exists i: \; B.i: \; R.i) \Rightarrow (\forall j: \; B.j: \; R.j)$
= {pred. calc.}
$(\forall i: \; B.i: \; \neg R.i) \lor (\forall j: \; B.j: \; R.j)$
= { $\lor$ distributes over $\forall$; unnesting}
$(\forall i,j: \; B.i \land B.j: \; \neg R.i \lor R.j)$
$\Leftarrow$ {since —Excluded Middle— $[i = j \Rightarrow \neg R.i \lor R.j]$ }
$(\forall i,j: \; B.i \land B.j: \; i = j)$.

(*End of Proof.*)

Furthermore, we have the simpler

(34) *Theorem* We have for any B and R

$$[(\exists i:: B.i) \equiv (\forall i: B.i: R.i) \Rightarrow (\exists i: B.i: R.i)]$$.

Proof We observe for any B and R

$(\forall i: B.i: R.i) \Rightarrow (\exists i: B.i: R.i)$
= {pred. calc. and de Morgan}
$(\exists i: B.i: \neg R.i) \vee (\exists i: B.i: R.i)$
= {combine the terms}
$(\exists i: B.i: \neg R.i \vee R.i)$
= {Excluded Middle and trading}
$(\exists i:: B.i)$.

(*End of Proof.*)

Corollary of (33) *and* (34)

(35) $[(\forall i,j: B.i \wedge B.j: i = j) \wedge (\exists i:: B.i) \Rightarrow$
        $((\forall i: B.i: R.i) \equiv (\exists i: B.i: R.i))]$.

Note that the antecedent equivales $(N i:: B.i) = 1$, i.e., $B.i$ holds for exactly one value of i .

(36) *Theorem* For any B and R

$$[(\forall i,j: B.i \wedge B.j: i = j)] \Rightarrow$$
$$[(\exists i:: B.i) \wedge (\forall i: B.i: R.i) \equiv (\exists i: B.i: R.i)] \wedge$$
$$[\neg(\exists i:: B.i) \vee (\exists i: B.i: R.i) \equiv (\forall i: B.i: R.i)]$$.

Proof We observe for any B and R

$[(\exists i:: B.i) \wedge (\forall i: B.i: R.i) \equiv (\exists i: B.i: R.i)]$
= {(34) and pred. calc.}
$[(\exists i: B.i: R.i) \wedge (\forall i: B.i: R.i) \equiv (\exists i: B.i: R.i)]$
⇐ {pred. calc. and (33)}
$[(\forall i,j: B.i \wedge B.j: i = j)]$,

thus establishing the first conjunct of (36); substition $R := \neg R$ and negating both sides yields the second conjunct.

(*End of Proof.*)

Let *DIF* be an alternative construct with disjoint guards, i.e., one for which the guards satisfy the antecedent of (36). With $R.i := wp.(S.i).X$, the first conjunct of (36) yields

(37) $[wp.DIF.X \equiv (\exists i: \ B.i: \ wp.(S.i).X)]$

on account of (29). On account of (27) and (26), the second conjunct of (36), with $R.i := wlp.(S.i).X$, yields

(38) $[wlp.DIF.X \equiv \neg BB \vee (\exists i: \ B.i: \ wlp.(S.i).X)]$.

From (37) it follows that *wp.DIF* inherits each disjunctivity property shared by all *wp.(S.i)* ; from (38) it follows that *wlp.DIF* enjoys, with the exception of universal disjunctivity, all disjunctivity properties shared by all *wlp.(S.i)* .

Now we are ready for our last theorem about the alternative construct.

(39) *Theorem* An alternative construct with disjoint guards is deterministic if all its statements are deterministic.

Proof Under the assumption of deterministic *S.i* we have to show

$$[wp.DIF.X \equiv (wlp.DIF)^*.X] \quad \text{for all } X \qquad .$$

To this end we observe for any *X*

$\quad (wlp.DIF)^*.X$
$= \quad \{\text{def. of the conjugate}\}$
$\quad \neg wlp.DIF.(\neg X)$
$= \quad \{(27) \text{ with } X := \neg X \ \}$
$\quad \neg(\forall i: \ B.i: \ wlp.(S.i).(\neg X))$
$= \quad \{\text{de Morgan}\}$
$\quad (\exists i: \ B.i: \ \neg wlp.(S.i).(\neg X))$
$= \quad \{\text{def. of the conjugate}\}$
$\quad (\exists i: \ B.i: \ (wlp.(S.i))^*.X)$
$= \quad \{ \ S.i \text{ is deterministic}\}$
$\quad (\exists i: \ B.i: \ wp.(S.i).X)$
$= \quad \{(37)\}$
$\quad wp.DIF.X \qquad .$

(End of Proof.)

The operational interpretation of the execution of *IF* is as follows. In those initial states in which none of the guards is satisfied, *IF* is semantically equivalent to *abort* . In those initial states in which at least one guard is satisfied, *S.i* is executed for a value of *i* such that *B.i* is initially satisfied.

The moral of our last theorem is that, as long as our programming language does not include nondeterministic basic statements such as *havoc* , each program is deterministic if, by definition, each alternative construct has disjoint guards. In FORTRAN this was guaranteed by admitting in a three-way test the guards $E > 0$, $E = 0$, and $E < 0$, respectively. In ALGOL 60 the disjointness was guaranteed by only admitting the guards B and $\neg B$. (We would render ALGOL 60's

$$\textbf{if } B \textbf{ then } S0 \textbf{ else } S1$$

by

$$\textbf{if } B \rightarrow S0 \;[]\; \neg B \rightarrow S1 \textbf{ fi} \qquad ,$$

and ALGOL 60's

$$\textbf{if } B \textbf{ then } S$$

by

$$\textbf{if } B \rightarrow S \;[]\; \neg B \rightarrow skip \textbf{ fi} \qquad .)$$

There were several reasons for the inclusion of nondeterminacy by means of nondisjoint guards.

The first reason was a fundamental one. The infix operator *max* being symmetrical, the assignment statement

$$z := x \ max \ y$$

is symmetrical in x and y . In the absence of the operator *max* , a regime like that of ALGOL 60 gives you two options

$$\textbf{if } x > y \textbf{ then } z := x \textbf{ else } z := y \qquad \text{and}$$
$$\textbf{if } x \geqslant y \textbf{ then } z := x \textbf{ else } z := y \qquad .$$

Both are asymmetric in x and y , and no methodology for program derivation can ever dictate the choice. By admitting nondisjoint guards it is possible to derive

$$\textbf{if } x \geqslant y \ \rightarrow \ z := x \;[]\; y \geqslant x \ \rightarrow \ z := y \textbf{ fi}$$

which —the [] being a symmetric separator— is symmetric in x and y .

The second reason was practical (or, if you prefer, opportunistic). From the task of designing operating systems we knew the necessity of being able to design nondeterministic programs. Facing the task of operating system design, we learned how to derive what we called at the time "synchronization conditions"; today we would call them "guards". By admitting nondisjoint guards, the methodology for deriving operating systems could be transferred, lock, stock, and barrel, to the derivation of sequential programs.

Remark   With the advent of the real-time interrupt, concurrency and nondeterminacy entered the world of programming at the same time and for many years they would remain closely associated notions. But they are very different: concurrency is an operational concept whereas nondeterminacy is not. The introduction of nondeterminacy into sequential programming helped in clarifying the distinction. (*End of Remark.*)

It was only at a subsequent stage that a further advantage of the inclusion of nondeterminacy was identified. It can offer an opportunity for separating the concerns for correctness from those for efficiency: sometimes it is relatively easy to design a correct nondeterministic algorithm, whose efficiency can subsequently be improved by subtly restricting its nondeterminacy (i.e., by strengthening guards).

$$* \quad * \quad *$$

So much for the semantics of straight-line programs. The term "straight-line" refers to the absence of repetition (or recursion); as long as we are restricted to straight-line programs, longer computations require longer program texts. We shall overcome this deficiency by introducing a repetitive construct; because the latter's semantics is most elegantly captured in terms of extreme solutions of equations in predicates, the next chapter is devoted to that topic.

CHAPTER 8

Equations in predicates and their extreme solutions

In the previous chapter we have encountered a number of statements S for which the predicate transformers $wlp.S$ and $wp.S$ were given in closed form. In the next chapter we shall encounter the statement DO , for which the predicates $wlp.DO.X$ and $wp.DO.X$ will be defined as solutions of equations of the form

$$(0) \qquad Y: \ [b.X.Y] \qquad .$$

Here, b is a predicate-valued function of two predicates, so $[b.X.Y]$ is a boolean scalar that for given X and Y is either *true* or *false* . In (0) we have followed our convention —here by the prefix "Y:"— of explicitly indicating the identity of the unknown(s), and thereby notationally distinguishing between the equation and the boolean expression that forms its body.

Remark We are aware of the fact that in our usage of the word "equation" we have generalized the traditional meaning "(Math.) statement of equality between two expressions (conveyed by the sign $=$)": little seems to be gained by writing

$$Y: \ ([b.X.Y] = true) \qquad .$$

More important than the syntactic requirement that "the sign $=$ " occurs, is that equations are equations in well-identified unknowns for which we may try to solve them: without the identification of the unknowns, we don't know what is meant by "the solutions of the equation". (*End of Remark*.)

Which predicates —if any— solve (0) depends (for given b) in general on which predicate we have chosen for X . We would like to consider the solution of (0) as a *function* of X —and that function is then the predicate transformer whose definition is our aim— . This goal requires that (0) has a *unique* solution for *each* X : if the solution is not unique, we don't define a function and if, for some X , it does not exist, we have failed to define a total function.

To begin, we focus our attention on uniqueness and existence of solutions. Because in this analysis it is irrelevant whether the equation has parameters like X , we consider the simpler equation

(1) $Y: [b.Y]$,

which may have any number of solutions. Next we consider (in terms of the same b) the equation

(2) $Y: ([b.Y] \land (\forall Z: [b.Z]: [Y \Rightarrow Z]))$,

which is, in general, much less tolerant than (1). We define predicate Q by

(3) $[Q \equiv (\forall Z: [b.Z]: Z)]$

and observe for any Y

$\quad (Y$ solves (2))
$= \quad${definition of (2)}
$\quad [b.Y] \land (\forall Z: [b.Z]: [Y \Rightarrow Z])$
$= \quad${interchange of universal quantifications}
$\quad [b.Y] \land [(\forall Z: [b.Z]: Y \Rightarrow Z)]$
$= \quad\{ \ Y \Rightarrow$ distributes over $\forall \ \}$
$\quad [b.Y] \land [Y \Rightarrow (\forall Z: [b.Z]: Z)]$
$= \quad\{(3)\}$
$\quad [b.Y] \land [Y \Rightarrow Q]$
$= \quad${from (3): $[b.Y] \Rightarrow [Y \Leftarrow Q]$ }
$\quad [b.Y] \land [Y \equiv Q]$
$= \quad${Leibniz}
$\quad [b.Q] \land [Y \equiv Q]$.

Comparing the first and last lines of the above, we see that (2) has at most one solution, viz., Q . If it exists, it solves $Y: [b.Y]$ as well and is called the latter's strongest solution, i.e.,

(4)   (Y is the strongest solution of $Y: [b.Y]$) $\equiv$ (Y solves (2))   .

And thus we have derived

(5) *Theorem* With $[Q \equiv (\forall Z: [b.Z]: Z)]$ we have

$$(Y \text{ is the strongest solution of } Y: [b.Y]) \equiv$$
$$[b.Q] \wedge [Y \equiv Q] \quad .$$

Existential quantification over Y of both sides yields

(6) *Theorem* With $[Q \equiv (\forall Z: [b.Z]: Z)]$ we have

$$(Y: [b.Y] \text{ has a strongest solution}) \equiv [b.Q] \quad .$$

Of these two theorems, the first one states that the strongest solution is unique if it exists (and in that case gives a closed expression for it), whereas the second one gives an expression for its existence.

Analogously to (2) and (4) we define in terms of equation

(7) $Y: ([c.Y] \wedge (\forall Z: [c.Z]: [Y \Leftarrow Z]))$

the notion of the weakest solution by

(8) $(Y \text{ is the weakest solution of } Y: [c.Y]) \equiv (Y \text{ solves (7)}) \quad .$

For b and c satisfying

(9) $(\forall Z: [c.Z \equiv b(\neg Z)])$.

we now observe

$(Y \text{ is the weakest solution of } Y: [c.Y])$ (*)
$= \quad \{(8) \text{ and } (7)\}$
$[c.Y] \wedge (\forall Z: [c.Z]: [Y \Leftarrow Z])$
$= \quad \{(9) \text{ and contra-positive}\}$
$[b.(\neg Y)] \wedge (\forall Z: [b.(\neg Z)]: [\neg Y \Rightarrow \neg Z])$
$= \quad \{\text{transforming the dummy: } Z := \neg Z \}$
$[b.(\neg Y)] \wedge (\forall Z: [b.Z]: [\neg Y \Rightarrow Z])$
$= \quad \{(2) \text{ and } (4)\}$
$(\neg Y \text{ is the strongest solution of } Y: [b.Y])$ (*)
$= \quad \{(5) \text{ with } Y := \neg Y \}$
$[b.Q] \wedge [\neg Y \equiv Q]$
$= \quad \{(9) \text{ and } (10)\}$
$[c.R] \wedge [Y \equiv R] \quad ,$

where we define R by

(10) $[R \equiv \neg Q] \quad .$

Furthermore, we observe

R

$=$ {(10) and (3)}

   $\neg(\forall Z:\ [b.Z]:\ Z)$

$=$ {de Morgan}

   $(\exists Z:\ [b.Z]:\ \neg Z)$

$=$ {transforming the dummy $Z := \neg Z$ }

   $(\exists Z:\ [b.(\neg Z)]:\ Z)$

$=$ {(9)}

   $(\exists Z:\ [c.Z]:\ Z)$.

Thus we have derived the duals of (5) and (6)

(11) *Theorem* With $[R \equiv (\exists Z:\ [c.Z]:\ Z)]$ we have

   (Y is the weakest solution of $Y:\ [c.Y]$) $\equiv$
   $[c.R] \wedge [Y \equiv R]$.

(12) *Theorem* With $[R \equiv (\exists Z:\ [c.Z]:\ Z)]$ we have

   ($Y:\ [c.Z]$ has a weakest solution) $\equiv [c.R]$.

En passant —see the lines marked (∗)— we have derived

(13)   (Y is the weakest solution of $Y:\ [c.Y]$) $\equiv$
   ($\neg Y$ is the strongest solution of $Y:\ [c.(\neg Y)]$)   .

The less specific term "extreme solution" is used for a strongest or a weakest one.

We now turn our attention to the existence of extreme solutions. We shall do so for equations of the special form

(14) $Y:\ [p.Y \Rightarrow q.Y]$.

To avoid duplication of work we exploit duality wherever we can. We specialize (13) to equations of the form of (14).

(15) *Theorem* We have for any predicate Z and predicate transformers p and q

   (Z is the strongest solution of $Y:\ [p.Y \Rightarrow q.Y]$) $\equiv$
   ($\neg Z$ is the weakest solution of $Y:\ [p^*.Y \Leftarrow q^*.Y]$)   .

Proof We observe for any Z , p , and q

(Z is the strongest solution of Y: $[p.Y \Rightarrow q.Y]$)
= {def. of conjugate, (6, 2)}
(Z is the strongest solution of Y: $[\neg p^*.(\neg Y) \Rightarrow \neg q^*.(\neg Y)]$)
= {(13) with $Y := \neg Z$ }
($\neg Z$ is the weakest solution of Y: $[\neg p^*.Y \Rightarrow \neg q^*.Y]$)
= {contra-positive}
($\neg Z$ is the weakest solution of Y: $[p^*.Y \Leftarrow q^*.Y]$)   .

(*End of Proof.*)

The above theorem allows us to restrict our attention and our proofs to strongest solutions; we shall quite often formulate the dual theorem as well.

(16) *Theorem* Consider equation

(17) Y: $[p.Y \Rightarrow q.Y]$.

Equation (17) has a strongest solution if p is monotonic and q is conjunctive over the solution set of (17). Equation (17) has a weakest solution if q is monotonic and p is disjunctive over the solution set of (17).

Proof We shall show the existence of a strongest solution for p monotonic and q conjunctive over the solution set of (17). On account of Theorem (6), our proof obligation is equivalent to showing that $(\forall Z: [p.Z \Rightarrow q.Z]: Z)$ solves (17). To this end we observe

$p.(\forall Z: [p.Z \Rightarrow q.Z]: Z)$
$\Rightarrow$ { p is monotonic, (5, 108)}
$(\forall Z: [p.Z \Rightarrow q.Z]: p.Z)$
$\Rightarrow$ {pred. calc.: $\forall$ is monotonic}
$(\forall Z: [p.Z \Rightarrow q.Z]: q.Z)$
= { q conjunctive over the solution set of (17)}
$q.(\forall Z: [p.Z \Rightarrow q.Z]: Z)$.

(*End of Proof.*)

In the above theorem, it is gratifying that the circumstances under which we can guarantee the existence of extreme solutions have been formulated in terms of monotonicity and junctivity: it is one more symptom of the significance of these concepts. The simplest way to meet the requirement of junctivity over the solution set is by taking a universally junctive function; the simplest universally junctive function is the identity function, and thus we are led to study, for monotonic f , the strongest solution of Y: $[f.Y \Rightarrow Y]$ and the weakest solution of Y: $[f.Y \Leftarrow Y]$. For reasons of duality —or, more precisely, on account of Theorem (15) and the fact that the conjugate of a monotonic predicate transformer is monotonic— we can confine our study to that of the strongest solution of Y: $[f.Y \Rightarrow Y]$.

For monotonic f , let Q be the strongest solution of

(18) Y: $[f.Y \Rightarrow Y]$;

from Theorem (16) we know that Q exists. From Theorem (5), we know that we can characterize Q by

$$[Q \equiv (\forall Z: [f.Z \Rightarrow Z]: Z)]   .$$

For formal proofs, however, it is often more convenient to characterize Q by its two salient properties, viz., that Q implies any solution, i.e.,

(19) $[f.Z \Rightarrow Z] \Rightarrow [Q \Rightarrow Z]$ for all Z

and that Q is a solution, i.e.,

(20) $[f.Q \Rightarrow Q]$

or, equivalently,

(20′) $[Z \Rightarrow f.Q] \Rightarrow [Z \Rightarrow Q]$ for all Z .

(The demonstration of the equivalence of (20) and (20′) is left as an exercise to the reader.) We shall illustrate this in the proof of the following theorem.

(21) *Theorem* Let f be a monotonic function from predicate pairs to predicates; for any X , let $g.X$ be the strongest solution of

(22) Y: $[f.(X,Y) \Rightarrow Y]$;

then g is a monotonic predicate transformer.

Proof Theorem (6, 57) states that, with the exception of universal junctivity, a function is as junctive in its components as it is in its total argument. Hence, f is monotonic in its first component and in its second component; from the latter and Theorem (16) we conclude that the strongest solution of (22) exists for any X and therefore that the introduction of the function g is justified (i.e., that $g.X$ is *uniquely* defined for *each* predicate X). We summarize our knowledge about g in analogy to (19) and (20′) by

(23) $[f.(X, Z) \Rightarrow Z] \Rightarrow [g.X \Rightarrow Z]$ for all X , Z

(24) $[Z \Rightarrow f.(X, g.X)] \Rightarrow [Z \Rightarrow g.X]$ for all X , Z .

The formal rendering of our proof obligation is

$$[P \Rightarrow Q] \Rightarrow [g.P \Rightarrow g.Q]   .$$

In order to meet it, we observe for any P , Q

$[g.P \Rightarrow g.Q]$

$\Leftarrow$ $\{(23)$ with $X,Z := P,g.Q$ $\}$

$[f.(P,g.Q) \Rightarrow g.Q]$

$\Leftarrow$ $\{(24)$ with $X,Z := Q,f.(P,g.Q)$ $\}$

$[f.(P,g.Q) \Rightarrow f.(Q,g.Q)]$

$\Leftarrow$ $\{$ f monotonic in its 1st argument$\}$

$[P \Rightarrow Q]$.

(End of Proof.)

A few remarks about the above proof are in order. Firstly, we observe that the first step exploits that the function applied to P is g , the middle step that the function applied to Q is g , and the last step that f is monotonic in its first component. Since all three data are needed, we may feel confident that our proof does not consist of more steps than necessary. Secondly, in view of the demonstrandum and the consequents of (23) and (24), it is clear that, for the g applied to P , property (24) is irrelevant and that therefore we *have* to use for that application of g that g enjoys property (23). Similarly, it is clear that we *have* to use the fact that the g applied to Q enjoys property (24). So, about the only freedom left in the design of the above proof is whether to appeal first to (23) or to (24). Quite often this turns out to be a minor choice in the sense that the proof can be completed in either case, be it at different expense. The guideline for this minor choice is the rule of thumb that, when working from consequent to antecedent —i.e., when allowing $\Leftarrow$ in the left-most column of our proof— we should appeal to formula (23) first. The reason is clear: an appeal to (23) duplicates the consequent, while an appeal to (24) complicates the consequent, and the rule of thumb suggests to try first to avoid the duplication of the complication.

It is instructive to see what would have happened, had we appealed to (24) first. We would have observed for any P and Q

$[g.P \Rightarrow g.Q]$.

$\Leftarrow$ $\{(24)$ with $X,Z := Q,g.P$ $\}$

$[g.P \Rightarrow f.(Q,g.Q)]$

$\Leftarrow$ $\{(23)$ with $X,Z := P,f.(Q,g.Q)$ $\}$

$[f.(P,f.(Q,g.Q)) \Rightarrow f.(Q,g.Q)]$

$\Leftarrow$ $\{$transitivity of $\Rightarrow$ $\}$

$[f.(P,f.(Q,g.Q)) \Rightarrow f.(P,g.Q)] \wedge [f.(P,g.Q) \Rightarrow f.(Q,g.Q)]$

$\Leftarrow$ $\{$ f monotonic in both its components$\}$

$[f.(Q,g.Q) \Rightarrow g.Q] \wedge [P \Rightarrow Q]$

$\Leftarrow$ $\{(24)$ with $X,Z := Q,f.(Q,g.Q)$ $\}$

$[P \Rightarrow Q]$.

In the above we needed a second appeal to (24) and, more serious, a second appeal to f's monotonicity in its second component, which was so far used only to demonstrate the existence of g . The extra price may be higher than we can afford.

The proof of Theorem (21) is a characteristic example of where to use the fact that a strongest solution implies all solutions —i.e., a formula like (23)— and of where to use the fact that a strongest solution is itself a solution —i.e., a formula like (24)— . Together with the rule of thumb these heuristics will guide us through the rest of this chapter.

In the case of monotonic f , our heuristics can be refined (and our proving power potentially enhanced), thanks to the fact that the extreme solutions turn out to solve a less tolerant equation as well. This is expressed by the beautiful theorem that in the oral tradition is known as "The Theorem of Knaster-Tarski". Because in this study of extreme solutions, our attention will be confined to monotonic functions, we shall first present

(25) *Theorem of Knaster-Tarski* For monotonic f

(26) $Y: [f.Y \equiv Y]$

has the same strongest solution as

(27) $Y: [f.Y \Rightarrow Y]$

and has the same weakest solution as

(28) $Y: [f.Y \Leftarrow Y]$.

Proof We can confine ourselves to demonstrating the existence and equality of the strongest solutions of (26) and (27), because existence and equality of the weakest solutions of (26) and (28) is the dual.

From Theorem (16), with f for p and the identity function (which is universally conjunctive) for q , we conclude that (27) has a strongest solution; let it be Q , i.e., we have

(29) $[f.Z \Rightarrow Z] \Rightarrow [Q \Rightarrow Z]$ for all Z

(30) $[f.Q \Rightarrow Q]$.

On account of these two properties of Q and the monotonicity of f , we now have to show that Q is also the strongest solution of (26), i.e., we have to establish

(31) $[f.Z \equiv Z] \Rightarrow [Q \Rightarrow Z]$ for all Z

(32) $[f.Q \equiv Q]$.

In order to establish (31), we observe for any Z

$$[Q \Rightarrow Z]$$
$\Leftarrow \quad \{(29)\}$
$$[f.Z \Rightarrow Z]$$
$\Leftarrow \quad \{\text{pred. calc.}\}$
$$[f.Z \equiv Z] \qquad .$$

In order to establish (32), we observe

$$[f.Q \equiv Q]$$
$= \quad \{\text{pred. calc.}\}$
$$[Q \Rightarrow f.Q] \wedge [f.Q \Rightarrow Q]$$
$= \quad \{(30)\}$
$$[Q \Rightarrow f.Q]$$
$\Leftarrow \quad \{(29) \text{ with } Z := f.Q \}$
$$[f.(f.Q) \Rightarrow f.Q]$$
$\Leftarrow \quad \{ f \text{ is monotonic}\}$
$$[f.Q \Rightarrow Q]$$
$= \quad \{(30)\}$
$$true \qquad .$$

(End of Proof.)

Notice that, formally, $(29) \Rightarrow (31)$ but $(30) \Leftarrow (32)$. The Theorem of Knaster-Tarski has two important methodological consequences. Formerly, the strongest solution Q of $Y: [f.Y \Rightarrow Y]$ was characterized by (29) and (30), and our heuristics told us where to appeal to which. We now see that, for monotonic f, we can replace each appeal to (30) by an appeal to the formally stronger (32). An added advantage is that the rule of thumb loses some of its significance: an early appeal to (30) might cause problems, but (32) always allows us to undo a premature replacement of Q by $f.Q$. The appeal to (29) —i.e., the stronger of (29) and (31)— or to (32) —i.e., the stronger of (30) and (32)— , when we have to *use* the fact that Q has been *defined* as a strongest solution, is the refinement of the heuristics alluded to before. This was the first methodological consequence of the Theorem of Knaster-Tarski.

The other methodological consequence follows from the fact that for monotonic f the strongest solution Q is equally well characterized by the two formally weaker (30) and (31)! To prove this claim, we have to demonstrate

$$[X \equiv Q]$$

by using

(33) $[f.X \Rightarrow X]$

(34) $[f.Z \equiv Z] \Rightarrow [X \Rightarrow Z]$ for all Z .

To this end we observe

$[Q \Rightarrow X]$
$\Leftarrow$ {(29) with $Z := X$ }
$[f.X \Rightarrow X]$
$=$ {(33)}
true   and

$[X \Rightarrow Q]$
$\Leftarrow$ {(34) with $Z := Q$ }
$[f.Q \equiv Q]$
$=$ {(32)}
true   .

The methodological impact of the above is that it suffices to demonstrate (33) and (34) when we have to show for monotonic f that some X defined elsewhere is the strongest solution of Y: $[f.Y \equiv Y]$.

Just to show how well these heuristics work, let us prove

(35) *Theorem* Let p and q be monotonic functions from predicate pairs to predicates; for any Y , let $f.Y$ be the strongest solution of

$$X: [p.(X,Y) \equiv X]   ;$$

for any X , let $g.X$ be the strongest solution of

$$Y: [q.(X,Y) \equiv Y]   .$$

Then the two equations

(36) $(X,Y): [(p.(X,Y), q.(X,Y)) \equiv (X,Y)]$

(37) $(X,Y): [(f.Y, g.X) \equiv (X,Y)]$

have the same strongest solution.

Proof We first verify the existence of the strongest solutions mentioned in the theorem. Because of Knaster-Tarski and the monotonicity of p and q , functions f and g are well-defined and moreover —Theorem (21)— monotonic. Pair-forming and selection being monotonic and functional composition preserving monotonicity, we now conclude that the strongest solutions of (36) and (37) exist.

Following the rules of the game, we formulate what is given about f and g by

(38) $[p.(X, Y) \Rightarrow X] \Rightarrow [f.Y \Rightarrow X]$ for all X , Y

(39) $[p.(f.Y, Y) \equiv f.Y]$ for all Y

(40) $[q.(X, Y) \Rightarrow Y] \Rightarrow [g.X \Rightarrow Y]$ for all X , Y

(41) $[q.(X, g.X) \equiv g.X]$ for all X .

In order to show that (36) and (37) have the same strongest solution, we define (P,Q) as the strongest solution of the one and show that the (P,Q) thus defined is the strongest solution of the other. Arbitrarily we choose to define (P,Q) as the strongest solution of (37) —and the reader is urged to verify that the other choice would have worked as well— . According to the rules of the game, we add to our data

(42) $[(f.Y , g.X) \Rightarrow (X,Y)] \Rightarrow [(P,Q) \Rightarrow (X,Y)]$ for all X , Y

(43) $[(f.Q , g.P) \equiv (P,Q)]$

and phrase our demonstrandum as

(44) $[(p.(X,Y) , q.(X,Y)) \equiv (X,Y)] \Rightarrow$
                                $[(P,Q) \Rightarrow (X,Y)]$ for all X , Y

(45) $[(p.(P,Q) , q.(P,Q)) \Rightarrow (P,Q)]$.

In order to demonstrate (44), we observe for any X , Y

$[(P,Q) \Rightarrow (X,Y)]$
$\Leftarrow$ $\{(42)\}$
$[(f.Y , g.X) \Rightarrow (X,Y)]$
$\Leftarrow \{(38)$ and $(40)\}$
$[(p.(X,Y) , q.(X,Y)) \Rightarrow (X,Y)]$
$\Leftarrow$ $\{$pred. calc.$\}$
$[(p.(X,Y) , q.(X,Y)) \equiv (X,Y)]$.

In order to demonstrate (45), we observe

$[(p.(P,Q) , q.(P,Q)) \Rightarrow (P,Q)]$
$=$ $\{(43)\}$
$[(p.(P,Q) , q.(P,Q)) \Rightarrow (f.Q , g.P)]$
$=$ $\{(39)$ with $Y := Q$; (41) with $X := P$ $\}$
$[(p.(P,Q) , q.(P,Q)) \Rightarrow (p.(f.Q , Q) , q.(P , g.P))]$
$\Leftarrow$ $\{$ p and q are monotonic$\}$
$[(P,Q) \Rightarrow (f.Q , g.P)]$
$=$ $\{(43)\}$
true   .

                                    (*End of Proof.*)

Remark Please note that the very last step would have been impossible, had (43) been formulated with $\Rightarrow$ instead of with $\equiv$. (*End of Remark.*)

Further remark Theorem (35) has been included in this chapter because we wanted to show in its proof our heuristics at work. For the sake of completeness, we would like to point out that the systematic application of the heuristics is not guaranteed to produce the shortest proof. For instance, (45) could have been demonstrated by observing

$$[(p.(P,Q)\,,\,q.(P,Q)) \Rightarrow (P,Q)]$$
$=$ {(43), four times}
$$[(p.(f.Q\,,\,Q)\,,\,q.(P\,,\,g.P)) \Rightarrow (f.Q\,,\,g.P)]$$
$=$ {(39) with $Y := Q$; (41) with $X := P$ }
    true   .

This shorter proof has the added attraction of not requiring a further appeal to the monotonicity of p and q , which could be regarded as a more fundamental reason for preferring it over our original demonstration. Its selective application of (43) in its first step, however, requires a noticeable amount of clairvoyance for its justification. (*End of Further remark.*)

After the above methodological interlude, we proceed with a more detailed study of the extreme solutions of

(46) $Y: [f.(X,Y) \equiv Y]$ with monotonic f .

According to Knaster-Tarski, its extreme solutions exist. We shall denote its strongest solution by $g.X$ and its weakest solution by $h.X$. Thanks to Knaster-Tarski, we can characterize these functions by

(47) $[f.(X,Z) \Rightarrow Z] \Rightarrow [g.X \Rightarrow Z]$ for all X , Z

(48) $[f.(X,g.X) \equiv g.X]$ for all X ,

(49) $[f.(X,Z) \Leftarrow Z] \Rightarrow [h.X \Leftarrow Z]$ for all X , Z ,

(50) $[f.(X,h.X) \equiv h.X]$ for all X .

We leave to the reader to derive from Knaster-Tarski and Theorem (21)

(51) *Theorem* Functions g and h are monotonic.

Now we are ready to demonstrate the beautiful

(52) *Theorem* Any type of conjunctivity enjoyed by f is enjoyed by h as well, and its dual

(52') *Theorem* Any type of disjunctivity enjoyed by f is enjoyed by g as well.

Remark Note that Theorem (51) is subsumed under Theorems (52); (51) has been derived separately because we need it for the proof of (52). (*End of Remark.*)

Proof In order to show that h is conjunctive over some V , we have to demonstrate

$$[h.(\forall X: \ X \in V: \ X) = (\forall X: \ X \in V: \ h.X)] \qquad .$$

In order to do so, we begin by observing —leaving for brevity's sake the range $X \in V$ implicitly understood—

$\quad [h.(\forall X:: \ X) \equiv (\forall X:: \ h.X)]$
$=\quad \{ \ [h.(\forall X:: \ X) \Rightarrow (\forall X:: \ h.X)]$ because —see (51)— h is monotonic, and (5, 108)$\}$
$\quad [h.(\forall X:: \ X) \Leftarrow (\forall X:: \ h.X)]$
$\Leftarrow\quad \{(49)$ with $X, Z := (\forall X:: \ X) , (\forall X:: \ h.X) \ \}$
$\quad [f.((\forall X:: \ X) , (\forall X:: \ h.X)) \Leftarrow (\forall X:: \ h.X)]$
$=\quad \{$quantification distributes over pair-forming$\}$
$\quad [f.(\forall X:: \ (X , h.X)) \Leftarrow (\forall X:: \ h.X)]$
$=\quad \{(50)\}$
$\quad [f.(\forall X:: \ (X , h.X)) \Leftarrow (\forall X:: \ f.(X , h.X))] \qquad .$

By now, it looks very much as if we have succeeded in reducing the conjunctivity of h to that of f . As a matter of fact, we almost have, but there is a minor complication: the universal quantifications in the implication at which we stopped are over predicates, whereas the definition of f's conjunctivity involves universal quantifications over predicate pairs. To overcome this complication, we construct a bag W of predicate pairs by

(53) $(X,Y) \in W \equiv X \in V \wedge [Y \equiv h.X]$

and observe that, thanks to the monotonicity of h , V and W are of the same junctivity type. Hence it suffices to show the implication at which we interrupted our calculation under the assumption that f is conjunctive over W . Giving the range of X explicitly, we present the final part of the proof:

$f.(\forall X:\ X \in V:\ (X, h.X))$
$=$ {one-point rule to introduce Y }
$f.(\forall X:\ X \in V:\ (\forall Y:\ [Y \equiv h.X]:\ (X,Y)))$
$=$ {unnesting}
$f.(\forall X,Y:\ X \in V\ \wedge\ [Y \equiv h.X]:\ (X,Y))$
$=$ {(53)}
$f.(\forall X,Y:\ (X,Y) \in W:\ (X,Y))$
$=$ { f assumed to be conjunctive over W }
$(\forall X,Y:\ (X,Y) \in W:\ f.(X,Y))$
$=$ {(53)}
$(\forall X,Y:\ X \in V\ \wedge\ [Y \equiv h.X]:\ f.(X,Y))$
$=$ {nesting}
$(\forall X:\ X \in V:\ (\forall Y:\ [Y \equiv h.X]:\ f.(X,Y)))$
$=$ {one-point rule to eliminate Y }
$(\forall X:\ X \in V:\ f.(X, h.X))$.

(End of Proof.)

Remark The last 7 steps of the above proof are not very exciting. The middle one does the work, the surrounding ones merely serve to introduce and to eliminate the names Y and W . Besides that, their sufficiency can be challenged, since we have silently identified

$$(\forall Z:\ Z \in W:\ f.Z)$$

with

$$(\forall X,Y:\ (X,Y) \in W:\ f.(X,Y))   .$$

(Using the functions "*left*" and "*right*" from predicate pairs to predicates, satisfying for any Z

$$[(left.Z\ ,\ right.Z) \equiv Z]   ,$$

the interested reader may prove the above by using three times the one-point rule.)

An alternative would have been to define W , using the "bagifier" **B** , by

$$W = (\mathbf{B}X:\ X \in V:\ (X, h.X))   ,$$

and to use rules for the manipulation of formulae containing such subexpressions. We did not include a separate subtheory of "bagification" because we would hardly use it in the rest of this little monograph. *(End of Remark.)*

The previous theorem dealt with the conjunctivity of the weakest solution of (46) and, by duality, with the disjunctivity of its strongest solution. We now turn to the conjunctivity of the strongest solution of (46) and, by duality, to the disjunctivity of its weakest solution. To this end we first prove

Theorem Let predicates X , Y satisfy

(54) $[f.(X,Y) \equiv Y]$;

then

(55) $[g.X \equiv g.true \wedge Y]$ for finitely conjunctive f

(55′) $[h.X \equiv h.false \vee Y]$ for finitely disjunctive f .

In other words: the strongest solution of (46) is for finitely conjunctive f the conjunction of an arbitrary solution of (46) and the constant predicate $g.true$, i.e., a predicate independent of X .

Proof We shall prove equivalence (55) by showing that each side implies the other.

(i)   We observe for any X , Y satisfying (54)

$\quad \lfloor g.X \Rightarrow g.true \wedge Y \rfloor$
$= \quad$ {predicate calculus}
$\quad [g.X \Rightarrow g.true] \wedge [g.X \Rightarrow Y]$
$\Leftarrow \quad$ { f , and hence g monotonic; (47) with $Z := Y$ }
$\quad [X \Rightarrow true] \wedge [f.(X,Y) \Rightarrow Y]$
$= \quad$ {predicate calculus and (54)}
$\quad$ *true*   .

(ii)   We observe for any X , Y satisfying (54)

$\quad [g.X \Leftarrow g.true \wedge Y]$
$= \quad$ {pred. calc., so as to tackle $g.true$ via (47)}
$\quad [g.true \Rightarrow g.X \vee \neg Y]$
$\Leftarrow \quad$ {(47) with $X, Z := true, g.X \vee \neg Y$ }
$\quad [f.(true, g.X \vee \neg Y) \Rightarrow g.X \vee \neg Y]$
$= \quad$ {pred. calc, preparing for f's conjunctivity}
$\quad [f.(true, g.X \vee \neg Y) \wedge Y \Rightarrow g.X]$
$= \quad$ {(54), the only thing given about Y }
$\quad [f.(true, g.X \vee \neg Y) \wedge f.(X,Y) \Rightarrow g.X]$
$= \quad$ { f is finitely conjunctive; pred. calc.}
$\quad [f.(X, g.X \wedge Y) \Rightarrow g.X]$
$= \quad$ {from (54) and (47) with $Z := Y$: $[g.X \wedge Y \equiv g.X]$ }
$\quad [f.(X, g.X) \Rightarrow g.X]$
$= \quad$ {(48)}
$\quad$ *true*   .

(*End of Proof.*)

Theorem (55) is most interesting for the weakest possible choice for Y , i.e., $h.X$. Thus we get the corollaries

(56) $[g.X \equiv g.true \wedge h.X]$ for finitely conjunctive f

(56') $[h.X \equiv h.false \vee g.X]$ for finitely disjunctive f .

From these and (52), we derive

(57) $[g.(X \wedge Y) \equiv g.X \wedge h.Y]$ for finitely conjunctive f

(57') $[h.(X \vee Y) \equiv h.X \vee g.Y]$ for finitely disjunctive f .

Proof We shall demonstrate (57). To this end, we observe for any X , Y

$\quad g.(X \wedge Y)$
$=\quad \{(56)$ with $X := X \wedge Y$ }
$\quad g.true \wedge h.(X \wedge Y)$
$=\quad \{ f$ being finitely conjunctive, so $-(52)-$ is h }
$\quad g.true \wedge h.X \wedge h.Y$
$=\quad \{(56)\}$
$\quad g.X \wedge h.Y$.

(*End of Proof.*)

Theorem (55) was primarily a stepping stone for theorems (56) and (57), which are the ones that are used. We shall use (56) to demonstrate

(58)   *Theorem* With the exception of universal conjunctivity and of *and-*continuity, the conjunctivity enjoyed by f is enjoyed by g as well,

and its dual

(58')   *Theorem* With the exception of universal disjunctivity and of *or-*continuity, the disjunctivity enjoyed by f is enjoyed by h as well.

Proof We demonstrate (58). For monotonic f , the monotonicity of g has been established in Theorem (51). For the remaining types of conjunctivity (positive, denumerable or finite), f is finitely conjunctive, and hence —see (56)—

$\quad\quad [g.X \equiv g.true \wedge h.X]$.

Because —see (52)— h is as conjunctive as f and the constant function —see (6, 37)— is positively conjunctive, g inherits —see (6, 41)— the three remaining conjunctivity types from f .

(*End of Proof.*)

After the above general theorems about the junctivity inheritance by extreme solutions we turn to the question under which circumstances we can give for extreme solutions closed expressions that are in general more pleasant to manipulate than expressions like the one given for Q in Theorem (5).

Theorem For monotonic f and any Y

(59) $[f.Y \Rightarrow Y] \Rightarrow [(\exists i: 0 \leqslant i: f^i.false) \Rightarrow Y]$

(59′) $[f.Y \Leftarrow Y] \Rightarrow [(\forall i: 0 \leqslant i: f^i.true) \Leftarrow Y]$.

This theorem is of interest because the quantified expressions in the consequents give us bounds on the extreme solutions of Y: $[f.Y \equiv Y]$, and, moreover, do so under the weakest junctivity assumption about f , viz., monotonicity.

Proof We shall confine ourselves to the demonstration of (59). Its consequent

$$[(\exists i: 0 \leqslant i: f^i.false) \Rightarrow Y]$$

is —see (5, 118)— equivalent to

$$(\forall i: 0 \leqslant i: [f^i.false \Rightarrow Y])   ,$$

and this proof obligation is amenable to mathematical induction over the natural numbers. With induction hypothesis

$$[f^i.false \Rightarrow Y]   ,$$

we observe

(i)   for the base

$[f^0.false \Rightarrow Y]$
=   {definition of functional iteration}
$[false \Rightarrow Y]$
=   {predicate calculus}
true   ,

(ii)   under the assumption of $[f.Y \Rightarrow Y]$, for the step

$[f^{i+1}.false \Rightarrow Y]$
$\Leftarrow$ {assumption about Y }
$[f^{i+1}.false \Rightarrow f.Y]$
=   {definition of functional iteration}
$[f.(f^i.false) \Rightarrow f.Y]$
$\Leftarrow$ { f is monotonic}
$[f^i.false \Rightarrow Y]$.

(*End of Proof.*)

From (59) we deduce that, for monotonic f , $(\exists i: 0 \leqslant i: f^i.false)$ would be the strongest solution of $Y: [f.Y \equiv Y]$ if it were to solve that equation. So let us investigate how we can establish

$$[f.(\exists i: 0 \leqslant i: f^i.false) \equiv (\exists i: 0 \leqslant i: f^i.false)] \qquad .$$

To this end we observe, starting at the right-hand side,

$(\exists i: 0 \leqslant i: f^i.false)$
$=$ {splitting the range}
$(\exists i: 0 = i: f^i.false) \vee (\exists i: 1 \leqslant i: f^i.false)$
$=$ {one-point rule; transforming the dummy}
$f^0.false \vee (\exists i: 0 \leqslant i: f^{i+1}.false)$
$=$ {definition of functional iteration; pred. calc.}
$(\exists i: 0 \leqslant i: f.(f^i.false))$
$=$ { f is denumerably disjunctive}
$f.(\exists i: 0 \leqslant i: f^i.false)$,

i.e., the assumption that f is denumerably disjunctive suffices. If, however, the predicates $f^i.false \ (0 \leqslant i)$ form a monotonic sequence, the weaker assumption that f is *or*-continuous suffices. We shall now show that, indeed, the sequence $f^i.false$ is weakening, by proving via mathematical induction over the natural numbers that for monotonic f

$$(\forall i: 0 \leqslant i: [f^i.false \Rightarrow f^{i+1}.false]) \qquad .$$

To this end we observe for monotonic f

(i) for the base

$[f^0.false \Rightarrow f^1.false]$
$=$ {def. of functional iteration}
$[false \Rightarrow f.false]$
$=$ {predicate calculus}
true ,

(ii) for the step

$[f^{i+1}.false \Rightarrow f^{i+2}.false]$
$=$ {def. of functional iteration}
$[f.(f^i.false) \Rightarrow f.(f^{i+1}.false)]$
$\Leftarrow$ { f is monotonic}
$[f^i.false \Rightarrow f^{i+1}.false]$.

Because continuity implies monotonicity, we have thus established

(60) *Theorem* For *or*-continuous f , the strongest solution of $Y: [f.Y \equiv Y]$ is $(\exists i: 0 \leqslant i: f^i.false)$,

whose dual is

(60') *Theorem* For *and*-continuous f , the weakest solution of $Y: [f.Y \equiv Y]$ is $(\forall i: \ 0 \leqslant i: \ f^i.true)$.

Theorem (60) raises our interest in the expressions $(\forall i: \ 0 \leqslant i: \ f^i.true)$ and $(\exists i: \ 0 \leqslant i: \ f^i.false)$. In view of later applications we shall evaluate them for an f of a special form.

Theorem Let, for finitely conjunctive p and some predicate X , f be defined by

$$[f.Y \equiv X \wedge p.Y] \qquad ;$$

then

(61) $[(\forall i: \ 0 \leqslant i: \ f^i.true) \equiv (\forall i: \ 0 \leqslant i: \ p^i.X)]$

(62) $[(\exists i: \ 0 \leqslant i: \ f^i.false) \equiv$
    $(\exists i: \ 0 \leqslant i: \ p^i.false) \wedge (\forall i: \ 0 \leqslant i: \ p^i.X)]$.

Proof We first prove by mathematical induction over the natural numbers that for each natural i

(∗) $[f^i.Y \equiv (\forall j: \ 0 \leqslant j < i: \ p^j.X) \wedge p^i.Y]$.

To this end we observe

(i) for the base

$\quad [f^0.Y = (\forall j: \ 0 \leqslant j < 0: \ p^j.X) \wedge p^0.Y]$
$= \quad \{\text{def. of functional iteration; } (5, 90)\}$
$\quad [Y \equiv true \wedge Y]$
$= \quad \{\text{pred. calc.}\}$
$\quad true \qquad ,$

(ii) for the step

$\quad f^{i+1}.Y$
$= \quad \{\text{def. of functional iteration}\}$
$\quad f.(f^i.Y)$
$= \quad \{\text{def. of } f \ \}$
$\quad X \wedge p.(f^i.Y)$
$= \quad \{\text{induction hypothesis } (\ast)\}$
$\quad X \wedge p.((\forall j: \ 0 \leqslant j < i: \ p^j.X) \wedge p^i.Y)$
$= \quad \{\text{def. of functional iteration; } p \ \text{finitely conjunctive}\}$
$\quad p^0.X \wedge (\forall j: \ 1 \leqslant j < i+1: \ p^j.X) \wedge p^{i+1}.Y$
$= \quad \{\text{pred. calc.}\}$
$\quad (\forall j: \ 0 \leqslant j < i+1: \ p^j.X) \wedge p^{i+1}.Y$.

Having thus established (∗), we demonstrate (61) by observing

$(\forall i:\ 0 \leqslant i:\ f^i.true)$
= $\quad \{(∗)$ with $Y := true \ \}$
$(\forall i:\ 0 \leqslant i:\ (\forall j:\ 0 \leqslant j < i:\ p^j.X) \wedge p^i.true)$
= $\quad \{$pred. calc.$\}$
$(\forall i:\ 0 \leqslant i:\ p^i.X \wedge p^i.true)$
= $\quad \{$because p is monotonic, so is $p^i \ \}$
$(\forall i:\ 0 \leqslant i:\ p^i.X)$.

In order to establish (62), we observe

$(\exists i:\ 0 \leqslant i:\ f^i.false)$
= $\quad \{(∗)$ with $Y := false \ \}$
$(\exists i:\ 0 \leqslant i:\ (\forall j:\ 0 \leqslant j < i:\ p^j.X) \wedge p^i.false)$
= $\quad \{$because for $j \geqslant i \ [false \Rightarrow p^{j-i}.X]$ and p^i is monotonic,
$\quad\quad [p^i.false \Rightarrow p^j.\ X]$ for $j \geqslant i \ \}$
$(\exists i:\ 0 \leqslant i:\ (\forall j:\ 0 \leqslant j:\ p^j.X) \wedge p^i.false)$
= $\quad \{\ \wedge$ distributes over $\exists \ \}$
$(\exists i:\ 0 \leqslant i:\ p^i.false) \wedge (\forall i:\ 0 \leqslant i:\ p^i.X)$.

$\hfill (End\ of\ Proof.)$

Because *and*-continuity of the above f is equivalent to *and*-continuity of p and —see (6, 24)— p being *and*-continuous and finitely conjunctive means that p is denumerably conjunctive, we can combine (60′) and (61) into

(63) *Theorem* For denumerably conjunctive p ,

$$(\forall i:\ 0 \leqslant i:\ p^i.X)$$

is the weakest solution of $Y:\ [X \wedge p.Y \equiv Y]$.

Theorems (60) and (62) yield

(64) *Theorem* For *or*-continuous and finitely conjunctive p ,

$$(\exists i:\ 0 \leqslant i:\ p^i.false) \wedge (\forall i:\ 0 \leqslant i:\ p^i.X)$$

is the strongest solution of $Y:\ [X \wedge p.Y \equiv Y]$.

Some justifying examples

Theorems (55) and (58) are less beautiful than theorem (52), which states that h inherits without constraints or exceptions the conjunctivity enjoyed by f . To show that the exceptions mentioned in (55) and (58) are not void —i.e., have not entered the picture merely as a result of our weakness as theorem provers— we shall construct some counter-examples.

Theorem (55) is restricted to finitely conjunctive f . We shall show that, for an f that is not finitely conjunctive, the conclusion $[g.X \equiv g.true \land Y]$ is, in general, invalid. Because all the other types of conjunctivity imply monotonicity, it suffices to come up with a monotonic —but not finitely conjunctive!— f , such that the conclusion of (55) does *not* hold.

The simplest choice for f that is monotonic but not finitely conjunctive is

$$[f.(X,Y) \equiv X \lor Y]$$.

With this choice, equation (46) becomes

$$Y\colon \ [X \lor Y \equiv Y]$$,

whose strongest solution $g.X$ is given by $[g.X \equiv X]$. Relation (54) becomes

$$[X \lor Y \equiv Y]$$,

which is satisfied by $[Y \equiv true]$. With these values for g and Y , the conclusion $[g.X \equiv g.true \land Y]$ of (55) would yield

$$[X \equiv true \land true]$$,

which is in general *false* . And this observation fully justifies (55)'s restriction to finitely conjunctive f .

Next we turn our attention to theorem (58), which excludes g's inheritance of universal conjunctivity and *and*-continuity.

To justify the exclusion of universal conjunctivity, we choose f given by

$$[f.(X,Y) \equiv Y]$$,

which is universally conjunctive. With this choice, equation (46) becomes

$$Y\colon \ [Y \equiv Y]$$,

whose strongest solution $g.X$ is given by $[g.X \equiv false]$. This g is definitely not universally conjunctive, and (58)'s exclusion of universal conjunctivity is therefore justified.

Finally, in order to show that theorem (58)'s exclusion of *and*-continuity is also justified, we need an f that is *and*-continuous, but *not* finitely conjunctive. (Otherwise —see (6, 24)— it would be denumerably conjunctive.) We cannot use our earlier f, given by $[f.(X,Y) \equiv X \vee Y]$, which is not finitely conjunctive, because the corresponding strongest solution is the identity function. But we can use it as a source of inspiration by considering

$$[f.(X,Y) \equiv X \vee p.Y]$$

with a carefully chosen p. We want p to be *and*-continuous, since that —see (6, 54)— will make f *and*-continuous. Furthermore, it would be nice if p were *or*-continuous as well, since that —see (60)— would give us a closed form for $g.X$. And we would like p to be so simple that there is hope of tackling that closed form analytically.

For predicates on a state space that has z as one of its integer coordinates we suggest for p the substitution $(z := z + 1)$, i.e., we propose for f

$$[f.(X,Y) \equiv X \vee (z := z + 1).Y] \qquad .$$

Because substitution is *and*-continuous, so is f. We are now considering the strongest solution $g.X$ of

$$Y: \ [X \vee (z := z + 1).Y \equiv Y] \qquad .$$

Because the substitution $(z := z + 1)$ is denumerably disjunctive, the dual of Theorem (63) tells us that $g.X$ is given by

$$[g.X \equiv (\exists i: \ 0 \leqslant i: \ (z := z + 1)^i.X)] \qquad ,$$

which can be simplified to

(∗) $[g.X \equiv (\exists i: \ 0 \leqslant i: \ (z := z + i).X)] \qquad .$

Our remaining obligation is to show that g, as given by (∗) is *not and*-continuous. To this end we consider the strengthening sequence $C.j \ (0 \leqslant j)$ given by $[C.j \equiv z \geqslant j]$. We observe, on the one hand,

$g.(\forall j: \ 0 \leqslant j: \ C.j)$
$= \quad$ {definition of C }
$g.(\forall j: \ 0 \leqslant j: \ z \geqslant j)$
$= \quad$ {consider $j = z + 1$ }
$g.false$
$= \quad$ {(∗)}
$(\exists i: \ 0 \leqslant i: \ (z := z + i).false)$
$= \quad$ {definition of substitution}
$(\exists i: \ 0 \leqslant i: \ false)$
$= \quad$ {pred. calc.}
$false \qquad .$

We observe, on the other hand,

$(\forall j\colon\ 0 \leqslant j\colon\ g.(C.j))$
$=$ {definition of C }
$(\forall j\colon\ 0 \leqslant j\colon\ g.(z \geqslant j))$
$=$ $\{(*)\}$
$(\forall j\colon\ 0 \leqslant j\colon\ (\exists i\colon\ 0 \leqslant i\colon\ (z := z + i).(z \geqslant j)))$
$=$ {definition of substitution}
$(\forall j\colon\ 0 \leqslant j\colon\ (\exists i\colon\ 0 \leqslant i\colon\ z + i \geqslant j))$
$=$ {consider $i = 0\ max\ (j - z)$ }
$(\forall j\colon\ 0 \leqslant j\colon\ true)$
$=$ {pred. calc.}
$true$.

Because $false \neq true$, g is not *and*-continuous and (58)'s exclusion of *and*-continuity is therefore justified.

And this concludes the construction of our third and last counter-example. We would like to add that the strengthening sequence $C.j$ that we used above is the standard vehicle for refuting *and*-continuity. This knowledge makes the choice of our last f less surprising.

CHAPTER 9

Semantics of repetitions

The reason for the inclusion of the previous chapter is that we shall use extreme solutions of equations in predicates to define the semantics of our next compound statement, known as the "repetition". We shall first study it in its simple form, in which it is composed of a guard B and a statement S. It is denoted by surrounding the guarded statement $B \to S$ by the special bracket pair $do \ldots od$. For the sake of brevity we shall call the resulting compound statement in this chapter "DO", i.e.,

$$DO = do \ B \to S \ od \qquad .$$

We shall define the semantics of DO in the usual way by defining (in terms of B and S) the predicate transformers $wlp.DO$ and $wp.DO$. But we shall define them differently. Were we to follow the same pattern as in the case of the straight-line programs, we would define (i) in some way the predicate transformer $wlp.DO$, (ii) in some way the predicate we would denote by $wp.DO.true$, and (iii) in terms of the previous two the predicate transformer $wp.DO$ by

(0) $\qquad [wp.DO.X \equiv wp.DO.true \land wlp.DO.X]$ for all X .

Because it is more convenient, in the case of the repetition we shall follow a slightly different route: we shall define $wlp.DO$ and $wp.DO$ independently and then prove (0) as a consequence of those definitions. Furthermore we shall show that $wlp.DO$ meets requirement R0 (universal conjunctivity) if $wlp.S$ does and that $wp.DO$ meets requirement R1 (excluded miracle) if $wp.S$ does.

170

We shall now define predicate transformers *wlp.DO* and *wp.DO* . We suggest that in this chapter the reader just accept these definitions as such, without wondering from where they come or what has inspired them. Such background information will be provided in the next chapter.

Predicate transformer *wlp.DO* is given by: for any predicate X , predicate *wlp.DO.X* is defined as the *weakest* solution of

(1) $Y: [(B \lor X) \land (\neg B \lor wlp.S.Y) \equiv Y]$.

Predicate transformer *wp.DO* is given by: for any predicate X , predicate *wp.DO.X* is defined as the *strongest* solution of

(2) $Y: [(B \lor X) \land (\neg B \lor wp.S.Y) \equiv Y]$.

Using
$$[wp.S.Y \equiv wp.S.true \land wlp.S.Y]$$

and distribution of $\lor$ over $\land$, we can rewrite the last equation as

$$Y: [(B \lor X) \land (\neg B \lor wp.S.true) \land (\neg B \lor wlp.S.Y) \equiv Y]$$.

With, for the same B and S , *IF* given by
$$IF = \textbf{if } B \to S \textbf{ fi}$$

we have —see (7, 27)—
$$[wlp.IF.Y \equiv \neg B \lor wlp.S.Y]$$.

Consequently, we can define *wlp.DO.X* as the weakest solution of

(3) $Y: [(B \lor X) \land wlp.IF.Y \equiv Y]$

and *wp.DO.X* as the strongest solution of

(4) $Y: [(B \lor X) \land (\neg B \lor wp.S.true) \land wlp.IF.Y \equiv Y]$.

These two equations are special instances of equation

(5) $Y: [Z \land wlp.IF.Y \equiv Y]$

and with this equation we start our analysis. Because *wlp.IF* is universally conjunctive, it is monotonic; since conjunction is also monotonic, the left-hand side of (5) is monotonic in Y and hence, according to Knaster-Tarski, the extreme solutions of (5) exist. Let $g.Z$ be the strongest solution of (5) and $h.Z$ its weakest solution.

Confronting (5) with (3) and (4), we see

(6) $[wlp.DO.X \equiv h.(B \lor X)]$

(7) $[wp.DO.X \equiv g.((B \lor X) \land (\neg B \lor wp.S.true))]$.

With these two relations we can prove (0) and can show that $wlp.DO$ meets requirement R0 . To begin with we observe that (i) Z is a universally conjunctive function of Z and (ii) $wlp.IF.Y$ is a universally conjunctive function of Y ; from these two observations we conclude on account of (6, 53)

$$Z \wedge wlp.IF.Y \text{ is a universally conjunctive function of } (Z,Y) \qquad .$$

From this we conclude on account of (8, 52)

(8) h is universally conjunctive

and, because universal conjunctivity implies finite conjunctivity, on account of (8, 57)

(9) $[g.(X \wedge Y) \equiv g.X \wedge h.Y]$.

In order to prove (0) we observe for any X

$wp.DO.true \wedge wlp.DO.X$
$= \quad \{(7) \text{ with } X := true \text{ and pred. calc.; (6)}\}$
$g.(\neg B \vee wp.S.true) \wedge h.(B \vee X)$
$= \quad \{(9) \text{ with } X,Y := (\neg B \vee wp.S.true), (B \vee X) \}$
$g.((B \vee X) \wedge (\neg B \vee wp.S.true))$
$= \quad \{(7)\}$
$wp.DO.X$.

Showing that requirement R0 is met means showing that $wlp.DO$ is universally conjunctive. From (6) we see

$$wlp.DO = h \circ p \text{ with } [p.X \equiv B \vee X] \text{ for all } X \qquad .$$

Because —see (6, 36)— the identity function is universally conjunctive, we conclude —see (6, 42)— that p is universally conjunctive. Because —see (8)— h is universally conjunctive and —see (6, 45)— functional composition is junctivity preserving, $wlp.DO$ (being $h \circ p$) is universally conjunctive. So we have dealt with R0 .

Showing that requirement R1 is met means showing

$$[wp.DO.false \equiv false] \qquad .$$

This is most easily shown by returning to the original definition (2) of $wp.DO$. The substitution $X := false$ in (2) defines $wp.DO.false$ as the strongest solution of

$$Y: [B \wedge wp.S.Y \equiv Y] \qquad ;$$

under the assumption that S meets requirement R1 , i.e., $[wp.S.false \equiv false]$, *false* is indeed the strongest solution of the above equation, and thus we have dealt with R1 .

From (0) and the universal conjunctivity of *wlp.DO* we conclude in the usual fashion —see (6, 37) and (6, 41)— that *wp.DO* is positively conjunctive.

In order to investigate the disjunctivity properties of *wlp.DO* and *wp.DO* , we first rewrite —on account of (5, 66)— (1) and (2) as

(10) Y: $[(\neg B \wedge X) \vee (B \wedge wlp.S.Y) \equiv Y]$ and

(11) Y: $[(\neg B \wedge X) \vee (B \wedge wp.S.Y) \equiv Y]$,

respectively. On account of the duals of (6, 36) and (6, 42):

- $(\neg B \wedge X)$ is a universally disjunctive function of X
- $(B \wedge wlp.S.Y)$ is as disjunctive a function of Y as $wlp.S.Y$
- $(B \wedge wp.S.Y)$ is as disjunctive a function of Y as $wp.S.Y$.

From the dual of (6, 53) we now conclude that, as functions of (X,Y) , the left-hand sides of (10) and (11) are as disjunctive as *wlp.S* and *wp.S* , respectively. From the inheritance theorems (8, 58') and (8, 52'), respectively, we now conclude

(12)   *Theorem*   With the exception of universal disjunctivity and *or-continuity*, *wlp.DO* inherits the disjunctivity properties of *wlp.S* .

(13)   *Theorem*   *wp.DO* inherits all disjunctivity properties of *wp.S* .

Finally, there is the question of *DO*'s determinacy, i.e., the question of whether

$$[(wlp.DO)^*.X \equiv wp.DO.X] \quad \text{for all } X \quad .$$

From the fact that *wlp.DO.X* is the weakest solution of (10) we derive —using de Morgan— that $(wlp.DO)^*.X$ is the strongest solution of

$$Y: [(B \vee X) \wedge (\neg B \vee (wlp.S)^*.Y) \equiv Y] \quad .$$

But if $[(wlp.S)^*.Y \equiv wp.S.Y]$ for any Y , i.e., if S is deterministic, this is the same equation as (2), the strongest solution of which has been defined to be *wp.DO.X* . Thus we have established

(14)   *Theorem*   For deterministic S , *DO* is deterministic.

As for the straight-line programs, we have now dealt for DO with the junctivity properties of its predicate transformers and with its determinacy. In the case of IF we had to deal with the additional complication that, even if the constituent statements were deterministic, IF could be nondeterministic; this complication could be dealt with by using plain predicate calculus: an alternative construct with mutually exclusive guards does not introduce nondeterminacy.

Our last theorem tells us that as far as the introduction of nondeterminacy is concerned, the repetition is an absolutely innocent statement. The repetition, however, raises a completely different issue, and that is the issue of guaranteed termination. Termination of DO is guaranteed for all initial states in which $wp.DO.true$ holds. According to (2) this predicate is given as the strongest solution of

$$Y: \quad [\neg B \lor wp.S.Y \equiv Y] \quad ,$$

but, for arbitrary S , that does not give us a sufficiently manageable expression for $wp.DO.true$. The risk of nontermination being introduced by the alternative construct depends on the value of $(\exists i:: B.i)$, i.e., it can be dealt with by the predicate calculus. In the case of the repetition, the question of guaranteed termination is a much more serious issue; in order to settle it, plain predicate calculus no longer suffices. It is this feature that makes the repetition intrinsically different from the straight-line programs, and this difference explains why we devote a whole separate chapter to it. Because the central notion that we need beyond the predicate calculus is that of well-founded sets, this is the place to insert the following intermezzo.

Intermezzo on well-founded sets

For a set D , a "relation from D to D" is a boolean function on the ordered pairs of elements from D . Relations are usually denoted by an infix operator; one of the best known relations is the equality relation: $x = y$ is identically $true$ if D is empty or a singleton set, it is not identically $true$ if D has more than one element.

We now consider a relation that we denote by an infix $<$, and pronounce as "less than".

Remark We denote our relation by an infix $<$ because that is the usual symbol in connection with well-founded sets. The choice is a little bit unfortunate because the relation need not be transitive (though in the most common well-founded sets it is). (*End of Remark.*)

Its most common model is that of a directed graph with the elements of D as its nodes and (the presence/absence of) an arrow from y to x as a coding for (the truth/falsity of) $x < y$.

In terms of this model, the fact that $<$ need not be transitive means that the presence of arrows from z to y and from y to x does not imply the presence of an arrow from z to x . Moreover, the graph is perfectly welcome to contain directed paths that are cyclic; in a moment it will, however, transpire that in such a case our attention will be focussed on acyclic subgraphs (where a subgraph is formed by the removal of a subset of nodes and all the arrows incident on removed nodes).

Let S be a subset of D ; the notion of "minimal element" is defined by

(15)   (x is a minimal element of S) $\equiv$
          $x \in S \wedge (\forall y: y < x: \neg y \in S)$.

Note that the minimal element need not be unique. Take for S the natural numbers; with the standard interpretation of $<$, 0 is the only minimal element of S , but with $<$ defined by

$$x < y \equiv (\exists p: p \text{ is a positive integer: } x + 2 \cdot p = y) \qquad ,$$

both 0 and 1 are minimal elements of S . Note also that a subset may have no minimal elements: take S empty, or take for S all integers and for $<$ the usual interpretation, or —in our graph model— take for S nodes on a cyclic path.

Well-foundedness of a subset C of D is defined by

(16)   (C is well-founded) $\equiv$
          $(\forall S:: (\exists x:: x \in C \cap S) \equiv (C \cap S \text{ has a minimal element}))$,

where, as in the rest of this intermezzo, the dummy S is understood to range over the subsets of D .

Remark In (16), dummy S further only occurs in the combination $C \cap S$. Indeed, well-foundedness of C is a statement about the subsets of C , viz., that their being non-empty equivales their having a minimal element. In view of what follows, $S \subseteq D$ is a more attractive range for S than $S \subseteq C$; hence the intersection with C . The intersection will be eliminated in a moment.
                                                          (*End of Remark.*)

Combining (15) and (16), we can now eliminate the notion of a minimal element:

(17)   (C is well-founded)
=   {(16)}
   $(\forall S:: (\exists x:: x \in C \cap S) \equiv (C \cap S$ has a minimal element))
=   {(15) with $S := C \cap S$ }
   $(\forall S:: (\exists x:: x \in C \cap S) \equiv$
        $(\exists x:: x \in C \cap S \wedge (\forall y: y < x: \neg(y \in C \cap S))))$
=   {definition of $\cap$; de Morgan}
   $(\forall S:: (\exists x:: x \in C \wedge x \in S) \equiv$
        $(\exists x:: x \in C \wedge x \in S \wedge (\forall y: y < x: \neg y \in C \vee \neg y \in S)))$
=   {trading}
   $(\forall S:: (\exists x: x \in C: x \in S) \equiv$
        $(\exists x: x \in C: x \in S \wedge (\forall y: y \in C \wedge y < x: \neg y \in S)))$.

For a while we leave our rephrasing of C's well-foundedness and introduce our next definition. (The high density of definitions is a bit annoying, but unavoidable because we want to establish the connections between existing pieces of mathematics.)

The validity of mathematical induction as a proof technique —we give the form known as "course-of-values induction"— is by definition given by

(18)   (mathematical induction over C is valid) $\equiv$
        $(\forall f:: (\forall x: x \in C: f.x) \equiv$
            $(\forall x: x \in C: f.x \Leftarrow (\forall y: y \in C \wedge y < x: f.y)))$,

where, as in the rest of this intermezzo, the dummy f is understood to range over the boolean functions on D . We are now free to establish a one-to-one correspondence between subsets S of D and boolean functions f on D by ruling that for each $x \in D$

(19) $f.x \equiv \neg x \in S$ or conversely $S = \{x | \neg f.x\}$.

Remark It was our desire to have f ranging over *all* functions of type $D \to$ bool that made us choose for S the range $S \subseteq D$.
(End of Remark.)

We now resume our calculation started at (17):

$(C$ is well-founded)
= {renaming dummy S according to (19)}
$(\forall f:: (\exists x: x \in C: \neg f.x) \equiv$
      $(\exists x: x \in C: \neg f.x \wedge (\forall y: y \in C \wedge y < x: f.y)))$
= {de Morgan and $\Leftarrow$ }
$(\forall f:: (\forall x: x \in C: f.x) \equiv$
      $(\forall x: x \in C: f.x \Leftarrow (\forall y: y \in C \wedge y < x: f.y)))$
= {(18)}

(mathematical induction over C is valid)   ,

i.e., we have established

(20) *Theorem* For a partially ordered set $(C, <)$

      $(C$ is well-founded) $\equiv$
      (mathematical induction over C is valid)   .

In other words, well-foundedness and validity of mathematical induction are just two sides of the same coin. The coin has a third side, which we can formulate after yet another definition, viz., the definition of a decreasing chain:

(21)   (elements $d.i$ $(0 \leqslant i)$ form a decreasing chain) $\equiv$
      $(\forall i: 0 \leqslant i: d.(i+1) < d.i)$.

Decreasing chains come in two kinds: infinite and finite chains. In the case of a finite chain, the dummy i in (21) should be properly bounded from above. The aforementioned third side of the coin is

(22) *Theorem* For a partially ordered set $(C, <)$

      $(C$ is well-founded) $\equiv$
      (all decreasing chains in C are finite)   .

Proof We rewrite the demonstrandum by negating both sides

      $(C$ is not well-founded) $\equiv$
      $(C$ contains an infinite decreasing chain)

and shall prove that each side implies the other. To this end we rewrite the left-hand side

$(C$ is not well-founded)

$=$ {see (17), with de Morgan and trading}

$(\exists S:: (\exists x: x \in C \cap S: true) \not\equiv$

$(\exists x: x \in C \cap S: (\forall y: y < x: \neg\, y \in C \cap S)))$

$=$ { $[LHS \Leftarrow RHS]$ and Golden Rule:

$[X \not\equiv Y \equiv X \wedge \neg Y \equiv X \Leftarrow Y]$ }

$(\exists S:: (\exists x: x \in C \cap S: true) \wedge$

$(\forall x: x \in C \cap S: (\exists y: y < x: y \in C \cap S)))$.

LHS $\Rightarrow$ *RHS*   Let S satisfy above conjunction. Then $C \cap S$ is non-empty and contains for every x a y that satisfies $y < x$; therefore $C \cap S$, and hence C , contains an infinite decreasing chain.

RHS $\Rightarrow$ *LHS*   Let C contain an infinite decreasing chain. Let B be the set of its elements; then

$(\exists x: x \in B: true) \wedge (\forall x: x \in B: (\exists y: y < x: y \in B))$.

Since, moreover, $B = C \cap B$, we can use B as instantiation for S .

(End of Proof.)

From theorems (20) and (22) we conclude

(23)   (mathematical induction over C is valid) $\equiv$
            (all decreasing chains in C are finite)   ,

a conclusion of particular interest for computing. Associating terminating repetitions with decreasing chains of finite length, we see some sort of link between terminating computations on the one hand and the validity of an inductive argument on the other hand: computation and mathematical induction go somehow hand in hand.

A well-known example of a well-founded set is presented by the natural numbers with $<$ in its standard meaning. The well-foundedness of the natural numbers is a postulate; we appeal to it whenever we prove something by mathematical induction over the natural numbers. With the rôle of C being played by the natural numbers, the set of all integers can correspond to D .

The natural numbers with $<$ form a very special well-founded set because it is totally ordered. A well-known example of a well-founded set with a partial order is provided by the sentences of a programming language —or any formalism with a similar grammar— , where " $<$ " is read as "occurs as subsentence in", which is clearly a transitive relation. The shrinking length of the subsequences ensures that decreasing chains are finite. It is this well-founded set that leads to "induction over the grammar" as we encountered in Chap. 7.

The above well-founded sets are still a little bit special in the sense that each element is the starting point of only a finite number of decreasing chains, whose lengths are therefore bounded. This is no longer necessarily true for what is known as the "lexical ordering". Lexical ordering is defined between equal-length sequences of elements from a (partially) ordered set. If the element ordering is well-founded and the sequences are of finite length, the lexical order makes the sequences a well-founded set. With x , y denoting elements and X , Y denoting sequences of the same length, the lexical order is defined by

$$xX < yY \;\equiv\; x < y \;\vee\; (x = y \;\wedge\; X < Y) \qquad .$$

(Here we did some overloading of the $<$.)

Example Consider the following one-person game. Given a finite bag of natural numbers; a move consists in replacing an x from the bag by the contents of a finite bag of natural numbers all $< x$. Show that the game terminates because the bag has become empty.

The argument is as follows. Let all numbers in the bag be $< n$; then this will remain so throughout the game. Characterize a bag by listing, for $n > j \geqslant 0$, the frequency with which j occurs in the bag. Each frequency being a natural number and the frequency listings being sequences of length n , the lexical order induces well-foundedness on the frequency listings; furthermore the frequency listing is lexically decreased by each move (because the numbers put into the bag are all smaller than the value removed). Successive frequency listings thus forming a decreasing chain from a well-founded set, this chain is finite, i.e., the game terminates. (*End of Example.*)

The above example clearly shows the sense in which (induction over) a well-founded set goes beyond (induction over) the natural numbers: the game is certain to terminate, but the initial bag does *not* define an upper bound on the number of moves that will finish the game. (Another way of buying that extension is the use of what is known as "transfinite induction", but we consider that price too high.)

In a moment the reader will encounter the central rôle of well-foundedness in all proofs of termination. This central rôle explains why computing scientists are so thoroughly familiar with the lexical order: it is their most beloved way of constructing more fancy well-founded sets from simpler ones.

(*End of Intermezzo on well-founded sets.*)

So far, the only structures we have worked with were boolean structures, which we called predicates. Because we need them for the formulation of our next theorem, the Main Repetition Theorem, we now turn our attention to structures of other types.

As source of inspiration we take another look at the by-now-familiar boolean structures and observe for any predicate X

> $true$
> $=$ {Excluded Middle}
> $[X \lor \lnot X]$
> $=$ {predicate calculus}
> $[(X \equiv true) \lor (X \equiv false)]$
> $=$ {(24) and predicate calculus}
> $[(\exists x: \ x \in \text{bool}: \ X = x)]$,

where the set bool is defined by

(24) $\text{bool} = \{true, false\}$.

Remark   In this discussion it seems unnecessarily pompous to make a notational distinction between a scalar type and the set of its possible values, so we won't do that. (*End of Remark.*)

Let D be some scalar domain (say, the set of integers). The above observation suggests how to define "a structure of type D", viz.,

(25) $(t \text{ is a structure of type } D) \equiv [(\exists x: \ x \in D: \ t = x)]$.

Notice that (trading and) application of the one-point rule yields

> $(t \text{ is a structure of type } D) \equiv [t \in D]$;

the last line may be read as "t is everywhere an element of D". With t a structure of type D and C a subset of D (say, C the natural numbers if D comprises the integers) $t \in C$ stands for a predicate that is not necessarily everywhere $true$.

Now we are ready to formulate the

(26) *Main Repetition Theorem*   Let $(D, <)$ be a partially ordered set;
let C be a subset of D such that $(C, <)$ is well-founded;
let statement S , predicates B and P , and structure t of type D satisfy

(27) $[P \land B \Rightarrow t \in C]$

(28) $(\forall x: \ x \in C: \ [P \land B \land t = x \ \Rightarrow \ wp.S.(P \land t < x)])$;

then, with $DO = \textbf{do }B \to S \textbf{ od}$,

(29) $[P \Rightarrow wp.DO.(P \wedge \neg B)]$.

(The well-informed reader will recognize in the above P an "invariant" of the repetition and in t a "variant function", which is the vehicle of the termination argument.)

Proof For the benefit of the designing programmer we have formulated the theorem in programming concepts such as wp and the statements S and DO . In order to prove the theorem, however, we shall first isolate its mathematical contents by eliminating from its formulation those program concepts.

Let us start with the demonstrandum (29) and rewrite it, by renaming the consequent, as

(30) $[P \Rightarrow Y]$,

and by asking ourselves what we (need to) know about Y in order to prove it. Well, Y has been defined as the strongest solution of an equation but, as it is the consequent of the demonstrandum, it can only matter that it is a solution. The latter does matter, for we have to take something about Y into account. Hence we shall use that Y satisfies —see also (48), a direct consequence of (11)—

(31) $[f.Y \equiv Y]$

with f , according to (11), for any Z given by

(32) $[f.Z \equiv (\neg B \wedge P) \vee (B \wedge wp.S.Z)]$

or, equivalently,

(33) $[f.Z \equiv (B \vee P) \wedge (\neg B \vee wp.S.Z)]$.

Because (31) embodies the only given fact about Y , the introduction of the named function f enables us to rephrase the question: what do we (need to) know about f ? In order to guarantee the existence of a Y satisfying (31) we probably need —on account of Knaster-Tarski—

(34) f is monotonic   ,

and our next question is what sufficient properties of f we can derive from the premises of the original theorem. This was the whole purpose of the introduction of the named function f . Its internal structure as given by (32) or (33) captures the semantics of the repetition; if we can use that internal structure to derive from the original premises some sufficient properties of

f , we get a new proof obligation which is simpler in the sense that the details of the semantics of the repetition have disappeared from the formulation.

Let us begin with premiss (28). We observe

$P \wedge B \wedge t = x \;\Rightarrow\; wp.S.(P \wedge t < x)$
$=$ {pred. calc., towards the 2nd disjunct of (32)}
  $P \wedge B \wedge t = x \;\Rightarrow\; B \wedge wp.S.(P \wedge t < x)$
$=$ {pred. calc., towards the 1st disjunct of (32)}
  $P \wedge (B \vee \neg P) \wedge t = x \;\Rightarrow\; B \wedge wp.S.(P \wedge t < x)$
$=$ {pred. calc.}
  $P \wedge t = x \;\Rightarrow\; (\neg B \wedge P) \vee (B \wedge wp.S.(P \wedge t < x))$
$=$ {(32) with $Z := P \wedge t < x$ }
  $P \wedge t = x \;\Rightarrow\; f.(P \wedge t < x)$.

Hence, (28) can be rephrased as

(35) $(\forall x: \; x \in C: \; [P \wedge t = x \;\Rightarrow\; f.(P \wedge t < x)])$.

Notice that now we have reached the stage that B occurs *only* in premiss (27) and the definition of f . So B has to be eliminated by confronting those two! To that end we observe

  (27)
$=$ {pred. calc., towards (33)}
  $[P \wedge \neg(t \in C) \Rightarrow (B \vee P) \wedge \neg B]$
$\Rightarrow$ {pred. calc., towards (33)}
  $(\forall Z:: \; [P \wedge \neg(t \in C) \Rightarrow (B \vee P) \wedge (\neg B \vee wp.S.Z)])$
$=$ {(33) to eliminate B }
  $(\forall Z:: \; [P \wedge \neg(t \in C) \Rightarrow f.Z])$.

Hence we suggest to replace premiss (27) by the seemingly weaker

(36) $(\forall Z:: \; [P \wedge \neg(t \in C) \Rightarrow f.Z])$.

(We said "seemingly", because the instantiation $Z := false$ shows that, thanks to the Excluded Miracle, (27) and (36) are, in fact, equivalent.)

Summarizing, we shall discharge the proof obligation of the Main Repetition Theorem by demonstrating

(i) $[P \Rightarrow Y]$ (30)

on account of

(ii)	$(C,<)$ is well-founded	
(iii)	$(\forall Z:: [P \wedge \neg(t \in C) \Rightarrow f.Z])$	(36)
(iv)	$(\forall x: x \in C: [P \wedge t = x \Rightarrow f.(P \wedge t < x)])$	(35)
(v)	$[f.Y \equiv Y]$	(31)
(vi)	f is monotonic .	(34)

We shall demonstrate (i) by showing separately

(37) $[P \wedge \neg(t \in C) \Rightarrow Y]$ and

(38) $[P \wedge t \in C \Rightarrow Y]$,

a case analysis that is strongly suggested by (iii).

Proof of (37) We observe

$[P \wedge \neg(t \in C) \Rightarrow Y]$
$=$ $\{(v)\}$
$[P \wedge \neg(t \in C) \Rightarrow f.Y]$
$=$ $\{(iii) \text{ with } Z := Y \}$
true   .

(End of Proof of (37).)

Proof of (38) As this is the part in which we have to exploit that $(C,<)$ is well-founded, we first massage our demonstrandum so as to make it amenable to a proof by mathematical induction over C . To this end we observe

$[P \wedge t \in C \Rightarrow Y]$
$=$ $\{$one-point rule$\}$
$[(\forall x: t = x: P \wedge x \in C \Rightarrow Y)]$
$=$ $\{$trading so as to make the range scalar$\}$
$[(\forall x: x \in C: P \wedge t = x \Rightarrow Y)]$
$=$ $\{$interchange of quantifications$\}$
$(\forall x: x \in C: [P \wedge t = x \Rightarrow Y])$.

In view of (ii), the latter is proved by deriving

(39) $[P \wedge t = x \Rightarrow Y]$ from

(40) $(\forall y: y \in C \wedge y < x: [P \wedge t = y \Rightarrow Y])$

for any x such that $x \in C$.

To this end we observe for any $x \in C$

(40)

= {interchange of quantifications}

$[(\forall y:\ y \in C \wedge y < x:\ P \wedge t = y\ \Rightarrow\ Y)]$

= {trading}

$[(\forall y:\ t = y:\ P \wedge y \in C \wedge y < x\ \Rightarrow\ Y)]$

= {one-point rule}

$[P \wedge t \in C \wedge t < x\ \Rightarrow\ Y]$

= {(37)}

$[P \wedge t < x\ \Rightarrow\ Y]$

$\Rightarrow$ {(vi)}

$[f.(P \wedge t < x) \Rightarrow f.Y]$

$\Rightarrow$ {(iv) and $x \in C$ }

$[P \wedge t = x\ \Rightarrow\ f.Y]$

= {(v)}

(39)   .

(*End of Proof of (38)*.)

(*End of Proof of (26)*.)

Acknowledgment In the proof of (38), the sequence "interchange of quantifications, trading, one-point rule" occurs twice. By a further parameterization of the argument, W. H. J. Feijen and A. J. M. van Gasteren have been able to avoid both that duplication and the case analysis (37) versus (38). Their proof is shorter than the one above and in a way quite elegant; we gave this proof because in the current state of our heuristic awareness we are unable to present their argument without pulling a sizeable rabbit out of the magician's hat. Besides their influence, we gratefully acknowledge that of C. A. R. Hoare and Lincoln A. Wallen in the formulation of the Main Repetition Theorem and the deduction of how to extract its mathematical contents. (The original proof, at the time of writing more than four years old, manipulated *wp* until the very end.) (*End of Acknowledgment.*)

The Main Repetition Theorem is by now decades old: to the best of our knowledge, R. W. Floyd was in the mid-sixties the first one to formulate it. Its first justification was in terms of a totally operational argument; since this was prior to any formal definition of the semantics of the repetition, an operational argument was the best that could be expected at the time. In the next decade, formal proofs were given under the constraint of the *or*-continuity of *wp.S* . Subsequent proofs that did not rely on *or*-continuity were so much more complicated that they strengthened the then-current opinion that it was wise to confine oneself to *or*-continuous *wp.S* . The significance of our proof is that it does not rely on *or*-continuity, is totally elementary, and is no more complicated than the original proofs that did rely on *or*-continuity. Its tentative moral is that the importance of *or*-continuity might have been overrated.

*   *   *

The Main Repetition Theorem involves well-founded sets because it deals with $wp.DO$, which captures guaranteed termination of the repetition. Since $wlp.DO$ is not concerned with guaranteed termination, we may expect $wlp.DO$ to be simpler to deal with than $wp.DO$. This expectation is confirmed by

(41) $[wlp.DO.X \equiv (\forall i: \ 0 \leqslant i: \ (wlp.IF)^i.(B \vee X))]$.

Proof We quote (6):

$$[wlp.DO.X \equiv h.(B \vee X)] \qquad ,$$

where $h.Z$ is the weakest solution of (5):

$$Y: \ [Z \wedge wlp.IF.Y \equiv Y] \qquad .$$

Because $wlp.IF$ is universally conjunctive, it is denumerably so, and, on account of (8, 63) with $X, p := Z, wlp.IF$, the weakest solution $h.Z$ of (5) is given by

(42) $[h.Z \equiv (\forall i: \ 0 \leqslant i: \ (wlp.IF)^i.Z)]$.

Substitution of (42) with $Z := B \vee X$ in (6) yields (41).

(*End of Proof.*)

Looking for closed expressions for $wp.DO.X$, we return to (2), which we rewrite as

(43) $Y: \ [k.Y \equiv Y]$ with k given by

(44) $[k.Y \equiv (B \vee X) \wedge (\neg B \vee wp.S.Y)]$

(in which k's hidden dependence on X should be noted).

(45) *Theorem* For *or*-continuous $wp.S$

$$[wp.DO.X \equiv (\exists i: \ 0 \leqslant i: \ k^i.false)] \qquad .$$

Proof By definition, $wp.DO.X$ is the strongest solution of (43). By (8, 60) that strongest solution is $(\exists i: \ 0 \leqslant i: \ k^i.false)$ if k is *or*-continuous. Since the constant function is *or*-continuous —see the dual of (6, 37)— and dis- and conjunction preserve *or*-continuity —see the duals of (6, 41) and (6, 43)— k is *or*-continuous if $wp.S$ is *or*-continuous.

(*End of Proof.*)

By rewriting (43) as

$$Y: \quad [k.true \wedge wlp.IF.Y \equiv Y]$$

we deduce from (8, 64) with $X, p := k.true, wlp.IF$

(46) *Theorem* For *or*-continuous $wlp.IF$

$$[wp.DO.X \equiv$$
$$(\exists i: 0 \leqslant i: (wlp.IF)^i.false) \wedge (\forall i: 0 \leqslant i: (wlp.IF)^i.(k.true))] ,$$

to the operational interpretation of which we shall return in the next chapter.

Both of the above two theorems have been used to prove the Main Repetition Theorem under the constraint of *or*-continuity. The point is that under that constraint the full generality of well-founded sets is not needed because induction over the natural numbers then suffices. Wherever in that case termination is guaranteed, it is also possible to state an upper bound on the number of iterations.

*   *
*

For the sake of completeness, we mention immediate consequences of (10) and (11):

(47) $[wlp.DO.X \equiv wlp.DO.(\neg B \wedge X)]$

(48) $[wp.DO.X \equiv wp.DO.(\neg B \wedge X)]$.

As consequences of the facts that $wlp.DO.X$ solves (10) and $wp.DO.X$ solves (11) we mention

(49) $[Q \Rightarrow \neg B] \Rightarrow [Q \wedge wlp.DO.X \equiv Q \wedge X]$

(50) $[Q \Rightarrow \neg B] \Rightarrow [Q \wedge wp.DO.X \equiv Q \wedge X]$,

which can be read as the statement that under validity of a Q that implies $\neg B$, DO acts as *skip*. We shall prove the latter one.

Proof We observe for any X and $[Q \Rightarrow \neg B]$

$\quad Q \wedge wp.DO.X$
$= \quad \{ wp.DO.X$ solves (11)$\}$
$\quad Q \wedge ((\neg B \wedge X) \vee (B \wedge wp.S.(wp.DO.X)))$
$= \quad \{ [Q \wedge \neg B \equiv Q]$ and $[Q \wedge B \equiv false] \}$
$\quad Q \wedge X$.

(End of Proof.)

In the same vein we derive

(51) *Theorem* With $DO = \textbf{do } B \to S \textbf{ od}$ and $DO' = \textbf{do } B' \to S' \textbf{ od}$,

$$DO;DO' = DO \quad \text{for } [B \Leftarrow B']\qquad .$$

Proof We observe for any X and $[B \Leftarrow B']$:

$wp."DO;DO'".X$
$=$ {definition of ; }
$wp.DO.(wp.DO'.X)$
$=$ {(48)}
$wp.DO.(\neg B \wedge wp.DO'.X)$
$=$ { $[\neg B \Rightarrow \neg B']$; (50) with $Q, B, DO := \neg B, B', DO'$ }
$wp.DO.(\neg B \wedge X)$
$=$ {(48)}
$wp.DO.X$.

For *wlp* the same argument applies.

(End of Proof.)

From (51) with $DO' := DO$ we derive

(52) $DO;DO = DO$,   i.e., predicate transformers
$wlp.DO$ and $wp.DO$ are idempotent   .

The proof of $DO = \textbf{do } B \to DO \textbf{ od}$ is left as an exercise to the reader. We are more interested in another exploration. We observe

$\neg B \vee wlp.IF.Y$
$=$ {definition of *wlp.IF* }
$\neg B \vee \neg B \vee wlp.S.Y$
$=$ {pred. calc.}
$\neg B \vee wlp.S.Y$

and

$\neg B \vee wp.IF.Y$
$=$ {definition of *wp.IF* }
$\neg B \vee (B \wedge (\neg B \vee wp.S.Y))$
$=$ {pred. calc.}
$\neg B \vee wp.S.Y$.

From these two equivalences, we conclude that equations (1) and (2) remain unchanged under the substitution $S := IF$; consequently, we have derived

$$DO = \textbf{do } B \rightarrow IF \textbf{ od} \qquad .$$

Because the above IF is an alternative with a single guarded command, B is the same predicate as BB (which was defined as the disjunction of the guards). Hence we may also write

(53) $DO = \textbf{do } BB \rightarrow IF \textbf{ od} \qquad .$

In the case of one guarded command, the transition from IF to DO , i.e., from **if** $B \rightarrow S$ **fi** to **do** $B \rightarrow S$ **od**, was notationally effectuated by replacing the parenthesis pair $\textbf{if} \cdots \textbf{fi}$ by the parenthesis pair $\textbf{do} \cdots \textbf{od}$, the two constructs semantically satisfying (53).

Our final linguistic proposal is to allow this transition also starting from an IF with more guarded commands, i.e., to

$IF = \textbf{if } B.0 \rightarrow S.0 \ \square \ B.1 \rightarrow S.1 \ \textbf{fi}$ corresponds

$DO = \textbf{do } B.0 \rightarrow S.0 \ \square \ B.1 \rightarrow S.1 \ \textbf{od}$, etc.

Each time, their semantics are coupled by (53) with $[BB \equiv (\exists i :: \ B.i)]$.

To conclude this chapter we shall show that the appropriate generalizations of equations (1) and (2) are

(54) $Y: \ [(BB \lor X) \land (\forall i: \ B.i: \ wlp.(S.i).Y) \equiv Y]$

(55) $Y: \ [(BB \lor X) \land (\forall i: \ B.i: \ wp.(S.i).Y) \equiv Y]$.

According to (53), the appropriate generalization of (1) is (1) with $B,S := BB,IF$, i.e.,

$$Y: \ [(BB \lor X) \land (\neg BB \lor wlp.IF.Y) \equiv Y] \qquad .$$

To show that this equation is the same as (54) we observe for the second conjunct

$\quad \neg BB \lor wlp.IF.Y$
$= \quad \{\text{def. of } BB \text{ and trading; def. of } wlp.IF \}$
$\quad (\forall i: \ B.i: \ false) \lor (\forall i: \ B.i: \ wlp.(S.i).Y)$
$= \quad \{ \ \forall \ \text{is monotonic}\}$
$\quad (\forall i: \ B.i: \ wlp.(S.i).Y) \qquad .$

The crucial step in establishing that (55) is the proper generalization of (2) is similarly

$\neg BB \lor wp.IF.Y$
=   {def. of $wp.IF$ }
   $\neg BB \lor (BB \land (\forall i: \;\; B.i: \;\; wp.(S.i).Y))$
=   {pred. calc.}
   $\neg BB \lor (\forall i: \;\; B.i: \;\; wp.(S.i).Y)$
=   {as above}
   $(\forall i: \;\; B.i: \;\; wp.(S.i).Y)$.

And this concludes the chapter on the semantics of the repetition.

CHAPTER 10

Operational considerations

In the two earlier chapters on semantics, we have defined the semantics of the compound statements $S0; S1$, IF , and DO by defining their predicate transformers in terms of the predicate transformers of their constituent statement(s). This raises the following question. What is or could be the relation between the execution of a compound statement and the executions of its constituent statements? Or, a bit more specifically, if we know how to implement the constituent statements, how could we then implement the semicolon, the alternative construct, and the repetition? These are the questions to which this chapter is devoted. We shall deal with the three compound statements in the order in which they have been introduced. (As is only to be expected, this is also the order of increasing complexity.)

In order to investigate possible implementations of the semicolon, let us define compound statement S by

$$S = S0; \ S1 \qquad .$$

We recall from (7, 23) and (7, 25), that the semantics of S is given by

$$[wlp.S.X \equiv wlp.S0.(wlp.S1.X)] \qquad \text{and}$$

$$[wp.S.X \equiv wp.S0.(wp.S1.X)] \qquad .$$

Looking at the right-hand side of, say, the latter one, we see that the predicate $(wp.S1.X)$ is —by virtue of the place where it occurs— a postcondition for $S0$ and —by virtue of its internal structure— a precondition for $S1$. We can do justice to this dual rôle of $(wp.S1.X)$ by identifying the state at which the execution of $S0$ terminates with the state in which the execution of $S1$ is

190

initiated. (Accordingly, in a computation in which the execution of $S0$ fails to terminate, the execution of $S1$ is not initiated at all.) This is the standard technique of implementing the semicolon: termination of its left-hand operand starts its right-hand operand. The technique is known as "sequential program execution". Barring a failure to terminate, the execution of "$S0$; $S1$;...; Sn" consists in executing, one after the other, the constituent statements in the order in which they occur in the sequence.

Remark   We do not take the position that the programmer prescribes computational behaviour; in particular, his use of the semicolon does not prescribe sequential program execution. As far as we are concerned, the executions of

$$\text{"}x := 2; y := 1\text{"} \qquad \text{and} \qquad \text{"}y := 1; x := 2\text{"}$$

could be identical. We prefer to view implementing the semicolon by means of sequential program execution as one of the implementor's options.
(End of Remark.)

In order to discuss the implementation of IF , we recall from (7, 27) and (7, 29)

$$[wlp.IF.X \equiv (\forall i: \quad B.i: \quad wlp.(S.i).X)]$$
$$[wp.IF.X \equiv BB \wedge (\forall i: \quad B.i: \quad wp.(S.i).X)] \qquad ,$$

where BB is given by

$$[BB \equiv (\exists i: \quad B.i: \quad true)] \qquad .$$

We distinguish two cases: whether or not BB holds in the initial state. We observe

$\qquad \neg BB$
$= \quad$ {definition of BB ; de Morgan}
$\qquad (\forall i: \quad B.i: \quad false)$
$\Rightarrow \quad$ {pred. calc.}
$\qquad (\forall i: \quad B.i: \quad wlp.(S.i).false)$
$= \quad$ {def. of $wlp.IF$ }
$\qquad wlp.IF.false \qquad .$

Because, for any statement S , $wlp.S.false$ holds in precisely those initial states for which each computation under control of S belongs to the class "eternal" —the class "finally $false$" being empty— , the above observation tells us how to implement *IF* in the case $\neg BB$: if initially all guards are *false* , the execution of *IF* should fail to terminate.

Next we consider the implementation of *IF* for an initial state in which *BB* holds. In that case, the implementation has to cater for the possibility that that initial state satisfies, for some postcondition X , the precondition *wp.IF.X* , in which case the execution of *IF* has to lead to a computation of the class "finally X" . We observe for any k

$$B.k \land wp.IF.X$$
$$= \quad \{\text{definition of } wp.IF \}$$
$$B.k \land BB \land (\forall i:\ B.i:\ wp.(S.i).X)$$
$$\Rightarrow \quad \{\text{pred. calc.}\}$$
$$wp.(S.k).X \qquad .$$

From this observation we conclude that the execution of *S.k* will do the job provided that the corresponding guard *B.k* holds in the initial state. The choice is, in general, indeed confined to the statement of a guarded command whose guard is initially *true* , because in general $\neg B.k \land wp.IF.X$ fails to imply $wp.(S.k).X$.

In summary: if all guards are *false* , execution of *IF* fails to terminate, otherwise it leads to an execution of one of the constituent statements whose guard is *true* . And this concludes our considerations about the implementation of the alternative construct.

We now turn our attention to the implementation of the repetition *DO* , given by

$$DO = \textbf{do}\ B \rightarrow S\ \textbf{od} \qquad ;$$

in the remainder of this chapter, *IF* will be used to denote —as in the previous chapter— the corresponding alternative construct, i.e.,

$$IF = \textbf{if}\ B \rightarrow S\ \textbf{fi} \qquad .$$

We recall from the previous chapter the definition of the semantics of *DO* , in particular from (9, 1) that, for any X , *wlp.DO.X* is the weakest solution of

(0) $Y:\ [(B \lor X) \land (\neg B \lor wlp.S.Y) \equiv Y]$

and from (9, 2) that, for any X , *wp.DO.X* is the strongest solution of

(1) $Y:\ [(B \lor X) \land (\neg B \lor wp.S.Y) \equiv Y]$.

We now observe for any X

 $true$

$=$ { $wlp.DO.X$ solves (0); $wp.DO.X$ solves (1)}
   $[(B \lor X) \land (\neg B \lor wlp.S.(wlp.DO.X)) \equiv wlp.DO.X] \land$
   $[(B \lor X) \land (\neg B \lor wp.S.(wp.DO.X)) \equiv wp.DO.X]$

$=$ {semantics of $if...fi$, $skip$, and semicolon}
   $[wlp."if \neg B \to skip \square B \to S; DO fi".X \equiv wlp.DO.X] \land$
   $[wp. "if \neg B \to skip \square B \to S; DO fi".X \equiv wp.DO.X]$.

Because the last result holds for all X and statements are semantically equivalent if they are characterized by the same predicate transformers, we have derived

(2) $if \neg B \to skip \square B \to S; DO fi = DO$.

With respect to the above equality, four remarks should be made.

The first remark is that substituting one side of (2) for the other is a standard practice in semantics-preserving program transformations; substituting the left-hand side for the right-hand side is called "unfolding", whereas the inverse substitution is called "folding".

The second remark is that, with our recently gained operational view of the alternative construct, we now see the standard implementation of the repetition: the computation consists of a (finite or infinite) sequence of executions of S —or, if you prefer, of IF — under the constraint that B holds at the initial state of each of these executions. When —and only when— B does *not* hold after a finite number (possibly zero) of these executions of S , the execution of DO terminates. To this operational interpretation of the repetition we shall return in a minute.

The third remark is that, viewed as an equation in DO , (2) may have more than one solution and, therefore, cannot serve as a definition of DO . Consider, for instance, the case $S = skip$; the equation

   $T: (if \neg B \to skip \square B \to T fi = T)$

has more than one solution: $T = skip$ is a solution, but so is $T = do B \to U od$ for any statement U .

The fourth remark is that, though we derived (2) from the semantic definitions (0) and (1), it happened in reality the other way round: (2) and its operational interpretation existed first, and it was (2) that inspired equations (0) and (1). Note that, in the derivation of (2), from the semantic definition of DO we used only that $wlp.DO.X$ and $wp.DO.X$ solve their respective equations; we did not use the fact that they have been defined as *extreme* solutions. The operational justification of the latter aspect of DO's semantic definition is one of the things we shall deal with in the remainder of this chapter.

With respect to a postcondition X , in Chap. 7, we partitioned computations into three mutually exclusive classes, viz., "finally X", "finally $\neg X$", and "eternal". Remember that, by definition, the class "finally *false*" is empty. We now apply this partitioning to the computations possible under control of DO .

Of these, the computations belonging to the class "finally X" are relatively simple. From our operational interpretation we see that, in the standard implementation of DO , a computation from the class "finally X" consists of a succession of a finite number (possibly zero) of terminating executions of IF , followed by a terminating execution of "*if* $\neg B \to skip$ *fi*" from the class "finally X". Hence it belongs to the class "finally $\neg B$" as well. This is fully in accordance with $[wlp.DO.(\neg B)]$, which follows from (9, 47) with $X := true$ and the universal conjunctivity of $wlp.DO$. So far, so good.

Next we ask ourselves, how under control of DO a computation from the class "eternal" can emerge. Analysis of the operational interpretation reveals that this can occur in the standard implementation of DO in two mutually exclusive ways; accordingly, for DO we shall partition the class "eternal" into two subclasses, which we shall denote by "outer eternal" and "inner eternal". They are defined as follows:

"outer eternal":   all eternal computations under control of DO that consist of an infinite sequence of executions of IF .

"inner eternal":   all eternal computations under control of DO that consist of a finite sequence of executions of IF .

An infinite sequence of executions of IF is possible only if each individual execution of IF —and in particular, each individual execution of S — terminates. The class "outer eternal" captures how the construct of the repetition may introduce the failure to terminate.

In the standard implementation of DO , a finite sequence of executions of IF gives rise to an eternal computation if and only if the execution of some IF —obviously the last one in the sequence— gives rise to an eternal computation under control of S ; this implies that that last IF has been initiated in a state satisfying $B \wedge \neg wp.S.true$. The class "inner eternal" captures how a repetition may fail to terminate due to nontermination of the constituent statement.

In order to connect our operational considerations with predicate transformers we recall —see Chap. 7— that for any S

$wp.S.true$:   holds in precisely those initial states for which no computation under control of S belongs to the class "eternal".

$wlp.S.(\neg X)$:   holds in precisely those initial states for which no computation under control of S belongs to the class "finally X".

With the aid of the conjugate, we can reformulate the above as

$(wp.S)^*.false$:   holds in precisely those initial states for which under control of S a computation belonging to the class "eternal" is possible.

$(wlp.S)^*.X$:   holds in precisely those initial states for which under control of S a computation belonging to the class "finally X" is possible.

The above tells us that $(wp.DO)^*.false$ states where "(outer or inner) eternal" is possible. We now turn to the more specific question of where a computation belonging to the class "outer eternal" is possible, i.e., the initial states for which the repetition mechanism as such may be responsible for the failure of termination. We characterize those initial states by the predicate C , i.e., we define

C: holds in precisely those initial states for which under control of DO a computation belonging to the class "outer eternal" is possible,

and our next task is to investigate whether we can now use our operational interpretation of DO to derive a formal characterization of C . Here we go.

We look at DO's unfolding

$$\textbf{if } \neg B \rightarrow skip \ \square \ B \rightarrow S; DO \textbf{ fi}$$

and conclude

(3) $[C \Rightarrow B]$

for, in each initial state satisfying C —see definitions of C and "outer eternal"— a terminating execution of IF is possible and hence B holds.

A second look at the unfolding tells us that it should be possible that, after the first execution of S , C holds again, i.e., that that first execution of S possibly belongs to the class "finally C". Hence we conclude

(4) $[C \Rightarrow (wlp.S)^*.C]$.

Combining (3) and (4) we conclude that C is a solution of

(5) $Y: [Y \Rightarrow B \wedge (wlp.S)^*.Y]$.

This equation has in general many solutions. We shall now show that C is its weakest solution.

Proof We discharge our proof obligation by showing $[Z \Rightarrow C]$ for any Z that solves (5).

Let Z be an arbitrary solution of (5). Then, from any initial state satisfying Z , a computation belonging to the class "finally Z" is possible under control of IF ; hence an infinite sequence of computations, each belonging to the class "finally Z" and each under control of IF , is possible under control of DO . Hence, for any initial state satisfying Z , a computation belonging to the class "outer eternal" is possible under control of DO , i.e., $[Z \Rightarrow C]$.

(End of Proof.)

Predicate C being the weakest solution of (5) is the same as $\neg C$ being the strongest solution of

$$Y: \quad [\neg B \vee wlp.S.Y \Rightarrow Y]$$

or —(7, 27), the semantics of IF — , equivalently, $\neg C$ being the strongest solution of

$$Y: \quad [wlp.IF.Y \Rightarrow Y]$$

or —monotonicity of $wlp.IF$ and (8, 25), the Theorem of Knaster-Tarski— $\neg C$ being the strongest solution of

(6) $Y: \quad [wlp.IF.Y \equiv Y]$.

But now we are on very familiar grounds. In the previous chapter we expressed —see (9, 6)— $wlp.DO$ in terms of h and —see (9, 7)— $wp.DO$ in terms of g , where for any Z , $h.Z$ and $g.Z$ were defined —see (9, 5)— as the weakest and the strongest solution of

(7) $Y: \quad [Z \wedge wlp.IF.Y \equiv Y]$.

The familiar ground is that equation (6) is a special instance of (7), viz., with $Z := true$. Thus we have derived $[\neg C \equiv g.true]$, and in connection with our definition of C :

(8) $g.true$:   holds in precisely those initial states for which no computation under control of DO belongs to the class "outer eternal".

We are very fortunate in having found an operational interpretation of $g.true$: because of

(9) $[g.Z \equiv g.true \wedge h.Z]$,

we now need only to be able to interpret $h.Z$ in order to be able to interpret both extreme solutions of (7). The operational interpretation of $h.Z$ is therefore our next concern. In fact we shall prove for any Z

(10) $h.Z$: holds in precisely those initial states for which in each computation under control of DO , Z holds prior to any execution of IF or of $if \neg B \rightarrow skip \ fi$.

Proof For some Z , let D hold in precisely those initial states for which, in each computation under control of DO , Z holds prior to each execution of IF or of $if \neg B \rightarrow skip \ fi$. We shall show $[D \equiv h.Z]$ by mutual implication.

$[D \Rightarrow h.Z]$

Since the initial validity of D implies, according to its definition, the initial validity of Z , we conclude

(11) $[D \Rightarrow Z]$.

Since under the initial validity of D , the computation under control of DO

- in the case $\neg B$ is empty,   and
- in the case B starts, on account of the definition of D , with an execution of S that belongs either to the class "eternal" or to the class "finally D ", we conclude —see definition of $wlp.S$ —

$$[D \Rightarrow \neg B \vee wlp.S.D]$$

or equivalently —see (7, 27)—

(12) $[D \Rightarrow wlp.IF.D]$.

Combining (11) and (12), we conclude that D is a solution of

(13) Y: $[Y \Rightarrow Z \wedge wlp.IF.Y]$.

Because —see (7) and (8, 25), the Theorem of Knaster-Tarski— $h.Z$ is the weakest solution of (13), we conclude $[D \Rightarrow h.Z]$.

$[D \Leftarrow h.Z]$

Let X be an arbitrary solution of (13). Then, from any initial state satisfying X , the computation under control of DO is empty or starts under control of IF with a computation that —because X solves (13)— belongs either to the class "eternal" or to the class "finally X ". Consequently, prior to each execution of IF or of $if \neg B \rightarrow skip \ fi$, X holds, and —again because X

solves (13)— Z holds as well. Thus each state satisfying X satisfies —on account of D's definition— D as well, i.e., $[D \Leftarrow X]$. Since this conclusion holds for any solution X of (13), we may instantiate it with the solution $h.Z$ for X , i.e., we have established $[D \Leftarrow h.Z]$.

(End of Proof.)

With the aid of (10) we can characterize the initial states for DO for which computations belonging to the class "inner eternal" are ruled out. Such a computation exists if and only if $B \wedge \neg wp.S.true$ holds prior to an execution of IF . Excluding the existence of such a computation therefore amounts to guaranteeing $\neg B \vee wp.S.true$ prior to each such execution, and from (10) we thus conclude

(14) $h.(\neg B \vee wp.S.true)$:
    holds in precisely those initial states for which no computation under control of DO belongs to the class "inner eternal".

We can similarly characterize the initial states for DO for which computations belonging to the class "finally $\neg X$" are ruled out. A computation belongs to the class "finally $\neg X$" if and only if $\neg B \wedge \neg X$ holds prior to the execution of *if* $\neg B \rightarrow skip$ *fi* . Excluding this possibility amounts to guaranteeing its negation prior to each execution of IF or of *if* $\neg B \rightarrow skip$ *fi* , and from (10) we therefore conclude

(15) $h.(B \vee X)$:
    holds in precisely those initial states for which no computation under control of DO belongs to the class "finally $\neg X$".

From this and the operational interpretation of $wlp.DO.X$ we conclude

$$[h.(B \vee X) \equiv wlp.DO.X] \quad \text{for any } X \quad ,$$

which —see (9, 6)— is in full accordance with our earlier definition of $wlp.DO.X$ as the *weakest* solution of (9, 1).

From (8) and (14) we conclude

$g.true \wedge h.(\neg B \vee wp.S.true)$:
    holds in precisely those initial states for which no computation under control of DO belongs to the class "outer eternal" or to the class "inner eternal".

Remembering that "outer eternal" and "inner eternal" form a partitioning of "eternal" and using (9), the above can be simplified to

$g.(\neg B \vee wp.S.true)$:
    holds in precisely those initial states for which no computation under control of DO belongs to the class "eternal".

From this and the operational interpretation of $wp.DO.true$ we conclude

$$[g.(\neg B \vee wp.S.true) \equiv wp.DO.true] \qquad ,$$

which —see (9, 7) with $X := true$— is in full accordance with our earlier definition of $wp.DO.true$ as the *strongest* solution of (9, 2) with $X := true$.

Though it is, strictly speaking, no longer necessary, we can combine the above operational interpretations of $h.(B \vee X)$ and $g.(\neg B \vee wp.S.true)$. Using (9, 9) this yields

$g.((B \vee X) \wedge (\neg B \vee wp.S.true))$:
    holds in precisely those initial states for which each computation under control of DO belongs to the class "finally X".

From this and the operational interpretation of $wp.DO.X$ we conclude that, for any X ,

$$[g.((B \vee X) \wedge (\neg B \vee wp.S.true)) \equiv wp.DO.X] \qquad ,$$

by which (9, 7) has been reconstructed in its full generality.

Remark The reader that is so inclined may verify statements like

$g.(B \vee X)$:   holds in precisely those initial states for which each computa-
             tion under control of DO belongs either to the class "finally
             X" or to the class "inner eternal",

etc.

                                                                             (*End of Remark.*)

$$* \qquad * \qquad *$$

In the previous chapter we promised to return in this chapter to the operational interpretation of theorem (9, 46), which states that, for *or-continuous* $wlp.IF$,

$$[wp.DO.X \equiv (\exists i:: (wlp.IF)^i.false) \wedge$$
$$(\forall i:: (wlp.IF)^i.((B \vee X) \wedge (\neg B \vee wp.S.true)))] \qquad .$$

Consider an initial state that satisfies the right-hand side. Because that initial state satisfies the universal quantification, no computation belongs —on account of $(B \vee X)$— to the class "finally $\neg X$" and no computation belongs —on account of $(\neg B \vee wp.S.true)$— to the class "inner eternal". Because that initial state satisfies the existential quantification, there exists for that initial state a sequence of IFs that would lead to a nonterminating computation, in particular to a nonterminating execution of one of the IFs . Because the class "inner eternal" is excluded, that IF fails to terminate on account of a precondition $\neg B$. Hence "outer eternal" is also excluded and "finally X" is guaranteed to occur.

The disadvantage of an operational argument like the above is that for the unwarned it is not patently obvious why the *or*-continuity of   *wlp.IF* is required for the validity of the above expression for *wp.DO.X* . The snag is that in the absence of *or*-continuity of *wlp.IF* , *wp.DO.X* may be satisfied by initial states for which we cannot construct a finite sequence of *IF*s long enough to guarantee the emergence of a precondition $\neg B$. The standard example is

$$
\begin{aligned}
&\textbf{\textit{do}}\ \ x > 0 \rightarrow x := x - 1 \\
&\quad \square\ \ x < 0 \rightarrow \text{``}x := \text{any natural number''} \\
&\textbf{\textit{od}}
\end{aligned}
$$

in which the semantics for "$x :=$ any natural number" is given by

$$
\begin{aligned}
&[wp.\text{``}x := \text{any natural number''}.\ (x \geqslant 0)]\ \wedge \\
&(\forall k :: [\neg wp.\ \text{``}x := \text{any natural number''}.\ (x < k)]])
\end{aligned}
$$

*   *   *

The reader will have noticed a difference in style between this chapter and most of the preceding ones. The underlying reason is that, in contrast to most of the preceding material, the operational interpretation forces us to mention individual program states explicitly. In all our theory about predicates, predicate transformers, and equations in predicates, *nowhere* did we need the existence of the individual states satisfying, for any predicate X , either X or $\neg X$: there was *no need at all* to introduce a predicate as a dichotomy of the state space, of which the individual states are the points. (In this sense our calculus of boolean structures can be viewed as a "pointless logic".) The other equally unattractive feature of this chapter is our frequent need to refer to computations. For good reasons we have made no effort to hide the relative clumsiness of operational arguments, conducted, as they always are, in terms of concepts that need not be mentioned.

CHAPTER 11

Converse predicate transformers

In Chap. 7, we took the decision to define programming language semantics in terms of weakest preconditions. In our last chapter we shall return to this decision, because we can also think of such things as "strongest postconditions". This chapter is devoted to the mathematical groundwork needed for that discussion.

Two predicate transformers f and k may be "each other's converses"; this state of affairs is denoted by $conv.(f,k)$, which expression is defined by

$$(0) \qquad conv.(f,k) \equiv (\forall X,Y:: \ [f.X \lor Y] \equiv [X \lor k.Y]) \qquad .$$

Predicate calculus suffices to establish

$$conv.(f,k) \equiv conv.(k,f) \qquad ,$$

which confirms that being each other's converses is, indeed, a symmetric relation.

Remark Note that we did not introduce an operator for "taking the converse of": we did not write —in analogy with the conjugate—

$$[k \equiv f^+] \qquad ,$$

nor did we render the symmetry by

$$[f^{++} \equiv f] \qquad .$$

The reason why we abstained from doing so is that the would-be expression could be meaningless, because —as the next theorem reveals— not every predicate transformer has a converse. (*End of Remark.*)

201

(1) *Theorem* For any f and k , the following three assertions are equivalent:

(i) $conv.(f,k)$

(ii)   (f is universally conjunctive) $\land$
        $(\forall Y:: [k.Y \equiv \neg(\forall X: [f.X \lor Y]: X)])$,

(iii) (f is monotonic) $\land$
        $(\forall Y:: [k.Y \equiv \neg(\text{the strongest solution of } X: [f.X \lor Y])])$.

Proof This theorem is proved by cyclic implication.

(i) $\Rightarrow$ (ii)

In order to use (i) to demonstrate f's universal conjunctivity, we observe for any bag V of predicates and any predicate Y , the range $X \in V$ for the dummy X being implicitly understood

$\quad [f.(\forall X:: X) \lor Y]$
$= \quad \{(i); (0) \text{ with } X := (\forall X:: X) \}$
$\quad [(\forall X:: X) \lor k.Y]$
$= \quad \{ \lor \text{ distributes over } \forall \}$
$\quad [(\forall X:: X \lor k.Y)]$
$= \quad \{\text{understood range is scalar}\}$
$\quad (\forall X:: [X \lor k.Y])$
$= \quad \{(i); (0)\}$
$\quad (\forall X:: [f.X \lor Y])$
$= \quad \{\text{understood range is scalar}\}$
$\quad [(\forall X:: f.X \lor Y)]$
$= \quad \{ \lor \text{ distributes over } \forall \}$
$\quad [(\forall X:: f.X) \lor Y]$.

Instantiating the above equivalence with $Y := \neg f.(\forall X:: X)$ and with $Y := \neg(\forall X:: f.X)$ proves

$$[f.(\forall X: X \in V: X) \equiv (\forall X: X \in V: f.X)]$$

by mutual implication.

In order to use (i) to demonstrate the second conjunct of (ii) we observe for any Y

$\quad \neg(\forall X: [f.X \lor Y]: X)$
$= \quad \{(i); (0)\}$
$\quad \neg(\forall X: [X \lor k.Y]: X)$
$= \quad \{(2) \text{ with } B := k.Y \}$
$\quad k.Y$.

Lemma For any predicate B

(2) $[\neg B \equiv (\forall X: [X \vee B]: X)]$.

Proof We observe for any B

   $[\neg B \Leftarrow (\forall X: [X \vee B]: X)]$
$\Leftarrow$ {instantiation with $X := \neg B$ }
   $[\neg B \vee B]$
$=$ {predicate calculus}
   true

and

   $[\neg B \Rightarrow (\forall X: [X \vee B]: X)]$
$=$ {predicate calculus}
   $[B \vee (\forall X: [X \vee B]: X)]$
$=$ { $\vee$ distributes over $\forall$ }
   $[(\forall X: [X \vee B]: B \vee X)]$
$=$ {range is scalar}
   $(\forall X: [X \vee B]: [B \vee X])$
$=$ {predicate calculus}
   true   .

(*End of Proof of (2).*)

(ii) $\Rightarrow$ (iii)

The first conjunct of (iii) follows trivially from the first conjunct of (ii).

To demonstrate the second conjunct of (iii) we first observe that equation $X: [f.X \vee Y]$ is the same equation as

(3) $X: [\neg Y \Rightarrow f.X]$.

With p given by $[p.X \equiv \neg Y]$ for all X , p is monotonic. For universally conjunctive f , f is conjunctive over the solution set of (3). On account of (8, 16) with $q := f$, we conclude that (3), and hence $X: [f.X \vee Y]$, has a strongest solution, which on account of (8, 5) equals $(\forall X: [f.X \vee Y]: X)$. The second conjunct of (iii) now follows from the second conjunct of (ii).

(iii) $\Rightarrow$ (i)

Rewriting the second conjunct of (iii) as

        $[\neg k.Y \equiv$ (the strongest solution of $X: [f.X \vee Y]$)]   , we can render it formally by the conjunction of

(4) $[f.(\neg k.Y) \lor Y]$ for all Y

(5) $[f.X \lor Y] \Rightarrow [\neg k.Y \Rightarrow X]$ for all X , Y .

The equivalence of (i) is proved by mutual implication. We observe for any X and Y

$[X \lor k.Y]$ $(*)$

=   {pred. calc.}

$[\neg k.Y \Rightarrow X]$

$\Leftarrow$ {(5)}

$[f.X \lor Y]$ $(*)$

$\Leftarrow$ {(4), rewritten as $[\neg f.(\neg k.Y) \Rightarrow Y]$ }

$[f.X \lor \neg f.(\neg k.Y)]$

=   {pred. calc.}

$[f.X \Leftarrow f.(\neg k.Y)]$

$\Leftarrow$ {(iii), f is monotonic}

$[X \Leftarrow \neg k.Y]$

=   {pred. calc.}

$[X \lor k.Y]$. $(*)$

Selecting the lines marked $(*)$, we see that we have shown under the assumption of (iii) that for any X , Y

$$[f.X \lor Y] \equiv [X \lor k.Y]   ,$$

i.e., (i).

(End of Proof of (1).)

As a corollary of Theorem (1) we see —(i) and (ii)— that a predicate transformer has a unique converse if it is universally conjunctive and has no converse otherwise.

We now derive some further theorems about the converse. We observe for any predicates X , Y

$[[X] \lor Y]$

=   {pred. calc.: $[X]$ is scalar}

$[X] \lor [Y]$

=   {pred. calc.: $[Y]$ is scalar}

$[X \lor [Y]]$,

in other words:

(6)   the "everywhere" operator is its own converse   .

In the model in which the predicates are boolean expressions in a bunch of variables and the "everywhere" operator is modelled by universal quantification over all those variables, the previous theorem is a special case of the following theorem. Here n plays the rôle of a variable, universal quantification over which could be included in the "everywhere" operator.

(7) $(\forall Zn:: \ [(\forall n:: \ Zn)] \equiv [Zn]) \equiv$
        $(\forall Xn, Yn:: \ [(\forall n:: \ Xn) \vee Yn] \equiv [Xn \vee (\forall n:: \ Yn)])$.

Here, the left-hand side captures n as one of the variables over which the "everywhere" operator quantifies universally, and the right-hand side expresses that, with f given by

$$[f.Zn \equiv (\forall n:: \ Zn)] \quad \text{for any } Zn \qquad ,$$

f is its own converse. For the names of the dummies we have chosen Xn , Yn , Zn to remind the reader that these predicates may punctually depend on the otherwise anonymous variable that we have named —viz., named "n" — so as to be able to denote universal quantification over it.

Proof The proof of (7) is by mutual implication.

$LHS \Rightarrow RHS$ We observe for any Xn , Yn

$\quad [(\forall n:: \ Xn) \vee Yn]$
$= \quad \{ \ LHS \ \text{with} \ Zn := (\forall n:: \ Xn) \vee Yn \ \}$
$\quad [(\forall n:: \ (\forall n:: \ Xn) \vee Yn)]$
$= \quad \{ \ n \ \text{is not a global variable of} \ (\forall n:: \ Xn) \ \}$
$\quad [(\forall n:: \ Xn) \vee (\forall n:: \ Yn)]$
$= \quad \{ \ n \ \text{is not a global variable of} \ (\forall n:: \ Yn) \ \}$
$\quad [(\forall n:: \ Xn \vee (\forall n:: \ Yn))]$
$= \quad \{ \ LHS \ \text{with} \ Zn := Xn \vee (\forall n:: \ Yn) \ \}$
$\quad [Yn \vee (\forall n:: \ Yn)]$

$LHS \Leftarrow RHS$ By instantiating RHS with $Yn := false$ (and renaming dummy Xn as Zn).

(End of Proof.)

Because for any B , X , Y

$$[(B \vee X) \vee Y] \equiv [X \vee (B \vee Y)] \qquad ,$$

we conclude

(8)   with f given by $[f.Z \equiv B \vee Z]$, f is its own converse   ,

and its corollaries — $B := true$ and $B := false$, respectively—

(9)   with f given by $[f.Z \equiv true]$, f is its own converse

(10)   the identity function is its own converse   .

Before tackling the problem of the converse of the substitution operator, we give a theorem about functional composition:

(11) $conv.(f0, k0) \wedge conv.(f1, k1) \Rightarrow conv.(f0 \circ f1, k1 \circ k0)$.

Proof We observe for any X , Y

$\quad [(f0 \circ f1).X \vee Y]$
$= \quad \{\text{def. of functional composition}\}$
$\quad [f0.(f1.X) \vee Y]$
$= \quad \{\ conv.(f0, k0)\ \}$
$\quad [f1.X \vee k0.Y]$
$= \quad \{\ conv.(f1, k1)\ \}$
$\quad [X \vee k1.(k0.Y)]$
$= \quad \{\text{def. of functional composition}\}$
$\quad [X \vee (k1 \circ k0).Y]$.

(*End of Proof.*)

We are now ready for the converse of the substitution operator. Recall from Chap. 6 the following: with n a variable, universal quantification over which is included in the "everywhere" operator, and E an expression of the same type, the substitution operator $(n := E)$ is a universally conjunctive predicate transformer. So it has a converse. Because —see (6, 67) and (6, 69)— the substitution operator $(n := E)$ is simpler if n is *not* a global variable of E , we shall deal with that case first.

We observe for predicates Xn , Yn , for n a variable over which the "everywhere" operator quantifies universally, and E an expression of which n is *not* a global variable

$\quad (n := E).Xn$
$= \quad \{(6, 67):\ n$ is not a global variable of $E\ \}$
$\quad (\forall n: n = E: Xn)$
$= \quad \{\text{trading}\}$
$\quad (\forall n:: n \neq E \vee Xn)$,

i.e., the substitution operator $(n := E)$ is the functional composition of universal quantification over n and disjunction with $n \neq E$. According to (7) and (8), respectively, these two predicate transformers are their own converses, and with (11) we have proved

(12) *Theorem* If n is *not* a global variable of E , the converse of the substitution operator $(n := E)$ is predicate transformer k , given by

$$[k.Yn \equiv n \neq E \lor (\forall n:: Yn)] \quad \text{for all } Yn \quad .$$

For a general expression En , of which n may be a global variable, the substitution operator $(n := En)$ —see $(6, 69)$— is more complicated; as to be expected, so is its converse.

(13) *Theorem* If n may be a global variable of En , the converse of the substitution $(n := En)$ is predicate transformer k , given with fresh variable y by

$$[k.Yn \equiv (\forall y: n = (n := y).En: (n := y).Yn)] \quad .$$

Proof In the following, x and y are fresh variables of the same type as n ; furthermore, universal quantification over x is deemed to be included in the "everywhere" operator. We observe for any state variable n , expression En of the same type, and predicates Xn , Yn

$\quad [(n := En).Xn \lor Yn]$
$=\quad \{ Xn \text{ independent of fresh } x \}$
$\quad [(x := En).((n := x).Xn) \lor Yn]$
$=\quad \{(11) \text{ and twice } (12), \text{ once with } n,E := x,En \text{ —and fresh } x \text{ is not a global variable of } En - \text{ and once with } n,E := n,x \text{ —and } n \text{ is not a global variable of fresh } x - \}$
$\quad [Xn \lor n \neq x \lor (\forall n:: x \neq En \lor (\forall x:: Yn))]$
$=\quad \{ Yn \text{ independent of } x \}$
$\quad [Xn \lor n \neq x \lor (\forall n:: x \neq En \lor Yn)]$
$=\quad \{\text{universal quantification over } x \text{ included in } [ ] \}$
$\quad [(\forall x:: Xn \lor n \neq x \lor (\forall n:: x \neq En \lor Yn))]$
$=\quad \{ Xn \text{ independent of } x , \lor \text{ distributes over } \forall \}$
$\quad [Xn \lor (\forall x:: n \neq x \lor (\forall n:: x \neq En \lor Yn))]$
$=\quad \{\text{trading, twice}\}$
$\quad [Xn \lor (\forall x: n = x: (\forall n: x = En: Yn))]$
$=\quad \{\text{renaming dummy } n \text{ by fresh } y \}$
$\quad [Xn \lor (\forall x: n = x: (\forall y: x = (n := y).En: (n := y).Yn))]$
$=\quad \{\text{one-point rule}\}$
$\quad [Xn \lor (\forall y: n = (n := y).En: (n := y). Yn)] \quad .$

Because in the first and last line of the above the "everywhere" operator is applied to operands that are independent of x , the proof has been completed.

(End of Proof.)

As a last theorem we mention

(14) *Theorem* Let W be a bag of predicate transformer pairs such that

(15) $(p,q) \in W \Rightarrow conv.(p,q)$;

then $conv.(f,k)$ holds for f , k , given by

$$[f.X \equiv (\forall p,q:\ (p,q) \in W:\ p.X)]\quad \text{for all}\ X$$

$$[k.Y \equiv (\forall p,q:\ (p,q) \in W:\ q.Y)]\quad \text{for all}\ Y\quad .$$

Proof Leaving for the dummies p , q the scalar range $(p,q) \in W$ implicitly understood, we observe for any X , Y

$\quad [f.X \vee Y]$

$= \quad \{\text{def. of } f \}$

$\quad [(\forall p,q::\ p.X) \vee Y]$

$= \quad \{\ \vee\ \text{distributes over } \forall \ \}$

$\quad [(\forall p,q::\ p.X \vee Y)]$

$= \quad \{\text{range is scalar}\}$

$\quad (\forall p,q::\ [p.X \vee Y])$

$= \quad \{(15)\}$

$\quad (\forall p,q::\ [X \vee q.Y])$

$= \quad \{\text{range is scalar}\}$

$\quad [(\forall p,q::\ X \vee q.Y)]$

$= \quad \{\ \vee\ \text{distributes over } \forall \ \}$

$\quad [X \vee (\forall p,q::\ q.Y)]$

$= \quad \{\text{def. of } k \}$

$\quad [X \vee k.Y]\qquad .$

(End of Proof.)

CHAPTER 12

The strongest postcondition

From the (rather operational) introduction of Chap. 7, we recall

$(wlp.S)^*.X$: holds in precisely those initial states for which there exists a computation under control of S that belongs to the class "finally X".

We also recall from that introduction the decision to use a predicate to introduce a dichotomy of the final state space. This led to a dichotomy of terminating computations —viz., a partitioning into the classes "finally X" and "finally $\neg X$"— and to the predicate transformers $wlp.S$ and $wp.S$, whose arguments are postconditions for S and whose values are preconditions for S.

Let us now pursue the use of a predicate to introduce a dichotomy of the initial state space and the corresponding dichotomy of computations (whether terminating or not), i.e., we propose to partition the computations into

"initially Y"   —all computations starting in an initial state satisfying Y
"initially $\neg Y$"   —all computations starting in an initial state satisfying $\neg Y$.

The intersection of the new class "initially Y" and the original class "finally X" is denoted by "initially Y & finally X". We conclude

$Y \wedge (wlp.S)^*.X$: holds in precisely those initial states for which there exists a computation under control of S that belongs to the class "initially Y & finally X".

209

Because each computation has an initial state, we conclude

(0) $[Y \wedge (wlp.S)^*.X \equiv false] \equiv$
         (no computation under control of S belongs to the class "initially
         Y & finally X")   .

This is a nice formula because its right-hand side is so symmetric in
"initially Y" and "finally X". Its left-hand side, being expressed in terms of
preconditions, is asymmetric in X and Y. This observation suggests to look
for an equivalent expression, similarly asymmetric in X and Y, but this
time expressed in terms of postconditions.

We introduce a predicate transformer $sp.S$ —the name "sp" coming
from "strongest postcondition"— , with an operational definition that is
completely analogous to the one we quoted above for $(wlp.S)^*$:

$sp.S.Y$:   holds in precisely those final states for which there exists a
             computation under control of S that belongs to the class "initially
             Y".

Therefore,

$X \wedge sp.S.Y$:   holds in precisely those final states for which there exists a
             computation under control of S that belongs to the class
             "initially Y & finally X",

and since each computation belonging to the class "initially Y & finally X"
has a final state, we conclude

(1) $[X \wedge sp.S.Y \equiv false] \equiv$
         (no computation under control of S belongs to the class "initially
         Y & finally X")   .

We now observe for any X, Y, S

$[wlp.S.X \vee Y]$
$=$ {de Morgan}
$[\neg wlp.S.X \wedge \neg Y \equiv false]$.
$=$ {def. of conjugate}
$[(wlp.S)^*.(\neg X) \wedge \neg Y \equiv false]$
$=$ {(0) with $X,Y := \neg X, \neg Y$ }
(no computation under control of S belongs to the class "initially $\neg Y$ &
finally $\neg X$")
$=$ {(1) with $X,Y := \neg X, \neg Y$ }
$[\neg X \wedge sp.S.(\neg Y) \equiv false]$
$=$ {de Morgan and def. of conjugate}
$[X \vee (sp.S)^*.Y]$,

and thus we have shown

(2) *Theorem* For all S , $conv.(wlp.S , (sp.S)^*)$.

Remark We recall the Hoare triple —see $(7, 3)$— :

$$\{Y\}\ S\ \{X\}   .$$

Two equivalent renderings are

(3) $[Y \Rightarrow wlp.S.X]$ and $[sp.S.Y \Rightarrow X]$.

They make clear that $wlp.S.X$ is the weakest solution of

(4) $Y:\ (\{Y\}\ S\ \{X\})$

and that $sp.S.Y$ is the strongest solution of

(5) $X:\ (\{Y\}\ S\ \{X\})$.

Furthermore —as the reader may verify by writing the implications in (3) as disjunctions— they confirm Theorem (2). (In passing, we would like to draw the reader's attention to the fact that the two —different!— equations (4) and (5) nicely illustrate the benefit of a notational convention that explicitly identifies the unknowns of equations.) (*End of Remark.*)

For the sake of completeness, we shall now determine the strongest postcondition $sp.S.Y$ for the statements S that we have introduced.

havoc

We recall $(7, 10)$

$$[wlp.havoc.X \equiv [X]]   \text{for all }\ X   .$$

Hence, on account of (2) and $(11, 6)$,

$$[(sp.havoc)^*.Y \equiv [Y]]   \text{for all }\ Y   ,$$

or, on account of $(6, 2)$,

(6) $[sp.havoc.Y \equiv \neg[\neg Y]]$ for all $Y.$

abort

We recall from $(7, 13)$

$$[wlp.abort.X \equiv true]   \text{for all }\ X   .$$

Hence, on account of (2) and (11, 9),

$$[(sp.abort)^*.Y \equiv true] \quad \text{for all} \quad Y$$

or, on account of (6, 2),

(7) $[sp.abort.Y \equiv false] \quad \text{for all} \quad Y$.

skip

We recall from (7, 16)

$$[wlp.skip.X \equiv X] \quad \text{for all} \quad X .$$

Hence, on account of (2) and (11, 10),

$$[(sp.skip)^*.Y \equiv Y] \quad \text{for all} \quad Y ,$$

or, on account of (6, 2),

(8) $[sp.skip.Y \equiv Y] \quad \text{for all} \quad Y$.

"n := E"

We recall from (7, 19)

$$[wlp."n := E".Xn \equiv (n := E).Xn] .$$

We first deal with the simple case in which n does *not* occur among the global variables of E . Then, on account of (2) and (11, 12),

$$[(sp."n := E")^*.Yn \equiv n \neq E \lor (\forall n:: Yn)] \quad \text{for all} \quad Yn ,$$

or, on account of (6, 2) and de Morgan,

(9) $[sp."n := E".Yn \equiv n = E \land (\exists n:: Yn)] \quad \text{for all} \quad Yn$.

In the case "$n := En$" , where n may occur among the global variables of En , we conclude, on account of (2) and (11, 13),

$$[(sp."n := En")^*.Yn \equiv$$
$$(\forall y: n = (n := y).En: (n := y).Yn)] \quad \text{for all} \quad Yn ,$$

or, on account of (6, 2) and the fact that substitution commutes with negation,

(10) $[sp."n := En".Yn \equiv$
$$(\exists y: n = (n := y).En: (n := y).Yn)] \quad \text{for all} \quad Yn ,$$
where y is a fresh variable.

"S0;S1"

We recall from (7, 23)

$$[wlp.``S0;S1".X \equiv wlp.S0.(wlp.S1.X)] \quad \text{for all } X \quad .$$

Hence, on account of (2) and (11, 11),

$$[(sp.``S0;S1")^*.Y \equiv (sp.S1)^*.((sp.S0)^*.Y)] \quad \text{for all } Y \quad ,$$

or, on account of (6, 2),

(11) $[sp.``S0;S1".Y \equiv sp.S1.(sp.S0.Y)] \quad \text{for all } Y \quad .$

IF

We recall from (7, 27)

$$[wlp.IF.X \equiv (\forall i: B.i: wlp.(S.i).X)] \quad \text{for all } X \quad ,$$

which, with

$$[f.i.X \equiv \neg B.i \vee X] \quad \text{for all } i, X \quad ,$$

can be rewritten as

$$[wlp.IF.X \equiv (\forall i:: (f.i).(wlp.(S.i).X))] \quad \text{for all } X \quad .$$

Hence, on account of (11, 8), (11, 11), (11, 14), and (2),

$$[(sp.IF)^*.Y \equiv (\forall i:: (sp.(S.i))^*.(\neg B.i \vee Y))] \quad \text{for all } Y \quad ,$$

or, on account of (6, 2) and de Morgan,

(12) $[sp.IF.Y \equiv (\exists i:: sp.(S.i).(B.i \wedge Y))] \quad \text{for all } Y \quad .$

DO

We recall from (9, 41)

$$[wlp.DO.X \equiv (\forall i: 0 \leqslant i: (wlp.IF)^i.(B \vee X))] \quad \text{for all } X \quad .$$

Hence, on account of (2), (11, 8), (11, 11), and (11, 14),

$$[(sp.DO)^*.Y \equiv (\forall i: 0 \leqslant i: B \vee ((sp.IF)^i)^*.Y)] \quad , \text{ or}$$

$$[(sp.DO)^*.Y \equiv B \vee (\forall i: 0 \leqslant i: ((sp.IF)^i)^*.Y)] \quad ,$$

or, on account of (6, 2) and de Morgan,

(13) $[sp.DO.Y \equiv \neg B \wedge (\exists i: 0 \leqslant i: (sp.IF)^i.Y)] \quad \text{for all } Y \quad .$

*   *   *

As it stands, (10) gives a rather complicated expression for $sp."n :=$
$En".Yn$. There is a very important case in which this expression can be
greatly simplified, viz., when the execution of the assignment statement does
not destroy information. For the assignment statement $"n := f.n"$ where
function f has an inverse, we observe for any Yn :

$$sp."n := f.n".Yn$$
$$= \quad \{(10)\text{ with } En := f.n \}$$
$$(\exists y: \ n = (n := y).(f.n): \ (n := y).Yn)$$
$$= \quad \{\text{def. of substitution}\}$$
$$(\exists y: \ n = f.y: \ (n := y).Yn)$$
$$= \quad \{\text{def. of } f^{-1} \}$$
$$(\exists y: \ y = f^{-1}.n: \ (n := y).Yn)$$
$$= \quad \{\text{one-point rule; } Yn \text{ independent of } y \}$$
$$(n := f^{-1}.n).Yn \quad .$$

In short

(14) $[sp."n := f.n".Yn \ \equiv \ (n := f^{-1}.n).Yn]$
        for any Yn and invertible f .

This is hardly more complicated than the Axiom of Assignment and —in
contrast to (10)— is regularly used in program development. The complexity
of (10), however, remains and is probably one of the reasons why a semantic
theory based on weakest preconditions turned out to be simpler than one
based on strongest postconditions.

In a moment of unwarranted optimism, one might try to save strongest
postconditions by restricting assignment statements of the form $n := f.n$ to
invertible f , but this would not work. No longer allowed to write

$$n := abs.n \quad ,$$

the programmer would write

$$\textbf{if } n \geqslant 0 \to skip \ [] \ n \leqslant 0 \to n := -n \ \textbf{fi} \quad ,$$

and the only consequence would be that the complication caused by
destruction of information would have been extended to the program text.

A further reason for rejecting the suggestion that assignment statements
should be information preserving in their execution is that it can be argued
that computer programs derive a major part of their utility from the fact that
their execution does destroy information by putting all inputs that produce
the same answer into the same equivalence class. Once that argument has
been bought, all justification for ruling out information destruction at the
lowest level seems to have evaporated.

The complicated form of the postcondition of the assignment statement is one reason for preferring a semantic theory based on preconditions. The other reason is that the strongest postcondition, being based on *wlp* , deals with partial correctness only. We could try to introduce a "strongest total postcondition" *stp.S.Y* by requiring for all X , Y

$$[Y \Rightarrow wp.S.X] \equiv [stp.S.Y \Rightarrow X] \qquad ,$$

but this effort would fail because *wp.S* is not necessarily universally conjunctive (as would be required by theorem (11, 1)). And since A. M. Turing we know —by now for more than half a century— that restricting ourselves to universally conjunctive *wp.S* , i.e., to programming languages in which each program is guaranteed to terminate, would amount to throwing away the baby with the bath water.

In short, by choosing the weakest precondition as the carrier for program semantics, we have been fortunate in being able to combine formal simplicity with the inclusion of the desired richness of our theory. And on this happy note, we conclude our final chapter and, thereby, this little monograph.

Index

216

Texts and Monographs in Computer Science

continued

David Gries
The Science of Programming
1981. XV, 366 pages

Micha Hofri
Probabilistic Analysis of Algorithms
1987. XV, 240 pages, 14 illus.

A.J. Kfoury, Robert N. Moll, and Michael A. Arbib
A Programming Approach to Computability
1982. VIII, 251 pages, 36 illus.

E.V. Krishnamurthy
Error-Free Polynomial Matrix Computations
1985. XV, 154 pages

Ernest G. Manes and Michael A. Arbib
Algebraic Approaches to Program Semantics
1986. XIII, 351 pages

Robert N. Moll, Michael A. Arbib, and A.J. Kfoury
An Introduction to Formal Language Theory
1988. X, 203 pages, 61 illus.

Franco P. Preparata and Michael Ian Shamos
Computational Geometry: An Introduction
1988. XII, 390 pages, 231 illus.

Brian Randell, Ed.
The Origins of Digital Computers: Selected Papers, 3rd Edition
1982. XVI, 580 pages, 126 illus.

Thomas W. Reps and Tim Teitelbaum
The Synthesizer Generator: A System for Constructing Language-Based Editors
1989. XIII, 317 pages, 75 illus.

Thomas W. Reps and Tim Teitelbaum
The Synthesizer Generator Reference Manual, 3rd Edition
1989. XI, 171 pages, 79 illus.

Arto Salomaa and Matti Soittola
Automata-Theoretic Aspects of Formal Power Series
1978. X, 171 pages

J.T. Schwartz, R.B.K. Dewar, E. Dubinsky, and E. Schonberg
Programming with Sets: An Introduction to SETL
1986. XV, 493 pages, 31 illus.

Texts and Monographs in Computer Science